CLOCK and CALENDAR
S K I L L S

Teaching Time to Special Students

Jean Bunnell
illustrated by
Scott W. Earle

J. WESTON
WALCH
PUBLISHER

Portland, Maine

Dedication

To my sister,

Nancy Carr,

with love

User's Guide to *Walch Reproducible Books*

As part of our general effort to provide educational materials which are as practical and economical as possible, we have designated this publication a "reproducible book." The designation means that purchase of the book includes purchase of the right to limited reproduction of all pages on which this symbol appears:

Here is the basic Walch policy: We grant to individual purchasers of this book the right to make sufficient copies of reproducible pages for use by all students of a single teacher. This permission is limited to a single teacher, and does not apply to entire schools or school systems, so institutions purchasing the book should pass the permission on to a single teacher. Copying of the book or its parts for resale is prohibited.

Any questions regarding this policy or requests to purchase further reproduction rights should be addressed to:

Permissions Editor
J. Weston Walch, Publisher
321 Valley Street • P. O. Box 658
Portland, Maine 04104-0658

1 2 3 4 5 6 7 8 9 10

ISBN 0-8251-2811-0

Copyright © 1987, 1996
J. Weston Walch, Publisher
P.O. Box 658 • Portland, Maine 04104-0658

Printed in the United States of America

Distributed By:
Grass Roots Press
Toll Free: 1-888-303-3213
Fax: (780) 413-6582
Web Site: www.literacyservices.com

Contents

Introduction .... *vii*
Of Clocks and Calendars .... *viii*

Large Clock Face 1
The Hour 3
    Worksheet 1 4
Half-Hour 5
    Worksheet 2 6
Hour/Half Past 7
    Worksheet 3 8
Quarter Past the Hour 9
    Worksheet 4 10
Quarter of the Hour 11
    Worksheet 5 12
Game: Draw the Time 13
Hour/Quarter Past/Half Past/Quarter of 19
    Worksheet 6 21
Five, Ten, Twenty, and Twenty-five Past 22
    Worksheet 7 24
Five, Ten, Twenty, and Twenty-five of 25
    Worksheet 8 26
Game: Time-O 27
Five, Ten, Twenty, and Twenty-five Past and Of 43
    Worksheet 9 44
Minutes Past and of the Hour 45
    Worksheet 10 46
Digital Time 47
    Worksheet 11 48
Digital Clocks 49
    Worksheet 12 50
Time in Words 51

Worksheet 13	52
Card Game: Matching the Time	53
Blank Clock Faces	59
Worksheet 14	60
Worksheet 15	61
Worksheet 16	62
Clocks without Numbers	63
Worksheet 17	64
Setting an Alarm	65
Worksheet 18	66
Using a Timer	67
Worksheet 19	69
A.M./P.M.	70
Worksheet 20	71
A.M./P.M. on Digital Clocks	72
Worksheet 21	73
Earlier/Later	74
Worksheet 22	75
Daily Schedule	76
Worksheet 23	77
TV Schedule	78
Worksheet 24	79
Worksheet 25	80
Worksheet 26	81
Bus Schedule	82
Worksheet 27	83
Board Game: On the Job	84
Game: The Print Shop	93
Days of the Week	98
Worksheet 28	99
Weekly Schedule	100
Worksheet 29	101
Game: A Busy Week	102
Months of the Year	114
Worksheet 30	115
Writing Dates with Numbers	116
Worksheet 31	117
Calendar	118

Holidays on Same Date	128
Worksheet 32	129
Holidays on Same Day	130
Worksheet 33	131
Personal Holidays	132
Worksheet 34	133
The Seasons	134
Worksheet 35	135
Seasonal Activities	136
Worksheet 36	137

Introduction

Clocks and calendars . . . they can provide order, helping people to work together and communicate. Or they can be mysterious jumbles of meaningless numbers. The worksheets, games, and activities in this book have been designed to "demystify" clocks and calendars for special students and to help them improve their time-telling skills.

The first section of worksheets introduces time-telling step by step, starting with the simplest concept of recognizing time on the hour. A large clock face is included for use as a visual aid in the classroom and can be used by students as a tactile learning experience.

Moving beyond the clock-related skills, students go on to distinguish between A.M. and P.M. and learn the days of the week and the months of the year. Pages are provided so students can make their own calendars, locating special days and holidays in each month.

Emphasis is placed on the practical applications of understanding time. Worksheets teach students to set an alarm clock and use a timer. Other worksheets help students read a TV schedule and decipher a bus schedule. With the current popularity of digital clocks, numerous activities stress recognizing the time on both a traditional clock face and in digital notation. In one game, students fill out a time card, adding units of time to achieve eight-hour days and forty-hour weeks. In another game, students complete a weekly schedule.

Teacher's material accompanies each worksheet and activity, listing specific objectives of the worksheet or activity and providing detailed suggestions for introducing the material. Games are included to provide fun and to motivate students to master their new skills.

Materials in this book will be useful in working with a large cross-section of students. Many of the worksheets dealing with both clocks and calendars could be used or easily adapted for use with nonreaders. Other worksheets present problem-solving activities for more able students.

This book presents a logical development of time-telling skills, but students do not necessarily have to complete all of the material in order. Teachers can choose the worksheets or activities that meet the needs of their students.

Of Clocks and Calendars

Telling the time with clocks and calendars is so much a part of our lives that most of us take it for granted. Yet clocks were not developed until the late 1200's. And the calendar used by most people in the Western world was not worked out by Pope Gregory until the 1580's.

The oldest known instruments designed for telling time were sundials, used more than 4,000 years ago. Sundials told the time my measuring the varying length of a shadow cast by the sun as it crossed the sky. Other early time-telling devices included hourglasses and water clocks. They worked by pouring either sand or water from one container into another at a steady rate. By measuring the amount of material poured, people could tell how much time had passed.

Earliest mechanical clocks had no hands or dial, but told the time by ringing a bell. The word *clock* probably comes from the French word *cloche* and the German word *Glocke*, both of which mean bell. By the mid-1300's, the dial and hour hand had been added. Minute and second hands became common by 1700.

Modern clocks range greatly in price, design, and accuracy. Most are either traditional dial clocks or digital clocks which became popular in the 1970's. Timekeeping in most clocks is determined by the frequency of some regularly repeating action, such as the swing of a pendulum. Clocks with stable frequencies keep time most accurately. Atomic clocks are the most accurate ever made. They are based on the vibrations of certain atoms or molecules, which almost always vibrate at the same rate per second. Thus, atomic clocks gain or lose only a few seconds in 100,000 years.

For ancient people, totally dependent on nature, the seeming rotation of the sun provided the most obvious unit of measuring time: the solar day. And the passing of the seasons dictated the length of another unit of time: the solar year. It was easy to see the changing position and shape of the moon so the lunar month became an intermediate measure of time.

We now know the time between successive full moons is about $29\frac{1}{2}$ days. Twelve lunar months would be about 354 days, almost 11 days shorter than the solar year of approximately $365\frac{1}{4}$ days. But a year of 13 lunar months would be almost 18 days longer than the solar year.

Most ancient calendars tried to compromise between the lunar and solar years, with some years of 12 months and some of 13. The Egyptians were probably the first people to adopt a predominantly solar calendar, recognizing a year of 365 days. The 12 months each had 30 days and an extra 5 days were added at the end of the year.

The Romans introduced a calendar borrowed from the Greeks in the mid-700's B.C. They recognized 10 months: Martius, Aprilis, Maius, Junius, Quintilis, Sextilis, September, October, November, and December. Their year had 304 days and seemed to ignore the remaining days, which fell in the middle of winter. Months were added, subtracted, and rearranged. By the first century B.C., the accumulated error caused by the inaccurate length of the year made the calendar about 3 months ahead of the seasons. While making numerous improvements in the calendar, Julius Caesar realigned the calendar with the seasons by ruling that the year we know as 46 B.C. should have 445 days. The Romans called it "the year of confusion."

On the advice of astronomers in 1582, Pope Gregory XIII made further corrections to the Julian calendar. The Gregorian calendar is so accurate that the difference between the calendar and the solar years is now only about 26 seconds.

Large Clock Face

Objective

    Make a large clock with hands that can be moved and set to different times.

Materials needed

    Copy of clock face on page 2
    Lightweight cardboard or oaktag (8"×11")
    Glue
    Scissors
    Acorn paper fastener

To make clock

1. Glue copy of clock face to cardboard. Spread glue over entire surface of paper.
2. Cut out clock face and the two hands.
3. Use acorn paper fastener to attach minute and hour hands to clock face.

Suggestions for use

1. Teacher can use clock as visual aid to introduce specific concepts.
2. Clock can be used by teacher or students to replicate a clock face on a worksheet as class is discussing the worksheet.
3. Students can make individual clocks to use for practice in telling time.
4. Clock can be used in game "Draw the Time" (see page 13).

Variation

    For some students, it may be difficult to distinguish between the shorter hour hand and the longer minute hand. It may be helpful to color the two hands different colors.

Name _____ Date _____

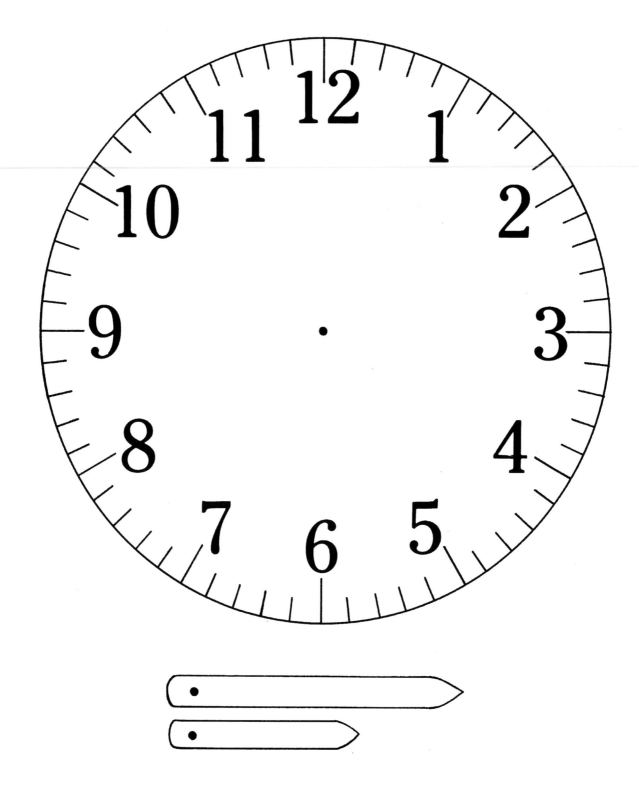

© 1987, 1996 J. Weston Walch, Publisher

Clock and Calendar Skills

THE HOUR
Directions for Worksheet 1

Objectives

1. Students distinguish between the hour hand and minute hand.
2. Students identify the time on the hour.
3. Students learn clock face, digital notation, and words for each time on the hour.

Materials needed

Large cardboard or oaktag clock face (see page 2)
Copy of Worksheet 1 for each student
Chalkboard and chalk
Pencil for each student

Introduce worksheet

1. Remove hands from large clock face.
2. Show clock face to students and ask what is missing (hands).
3. Hold up clock hands. The first hand to talk about is the shorter hand. Attach it to the clock face.
4. This hand is the hour hand. It tells what hour we are talking about. When it is near the 3 (turn hand to 3), it is close to 3 o'clock. When it is near the 7 (turn hand to 7), it is close to 7 o'clock.
5. The minute hand is longer than the hour hand. It helps us know exactly what time it is. (Attach minute hand to clock.)
6. When the minute hand is on 12, we know it is exactly the hour that the little hand is pointing to.
7. If the hour hand is on 3 and the minute hand is on 12 (arrange hands), it is exactly 3 o'clock—not before or after.
8. We can write that in digits—3:00. (Write it on the chalkboard.)
9. Or we can write it in words—three o'clock. (Write it on the chalkboard.)
10. Continue with several examples for the class. Use the cardboard clock to show time on the hour. Write the corresponding time (in digits and in words) on the chalkboard.
11. Students complete the worksheet by writing the indicated time on each clock face in digits and in words.

Suggestion

If students are unsure how to spell each of the numbers (one, two, etc.), write the numbers on the board so students can refer to them while doing the worksheet.

Name _____ Date _____

Reproducible Worksheet 1

What time is shown on each clock face?
Write the time in digits and in words.

_____ :00
_____ o'clock

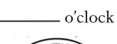

_____ :00
_____ o'clock

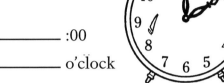

_____ :00
_____ o'clock

_____ :00
_____ o'clock

_____ :00
_____ o'clock

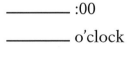

_____ :00
_____ o'clock

_____ :00
_____ o'clock

_____ :00
_____ o'clock

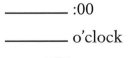
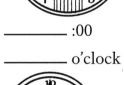
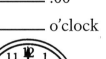

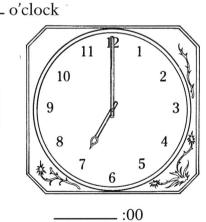
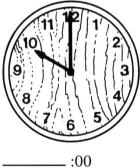
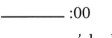

_____ :00
_____ o'clock

_____ :00
_____ o'clock

_____ :00
_____ o'clock

_____ :00
_____ o'clock

Clock and Calendar Skills

HALF HOUR
Directions for Worksheet 2

Objectives

1. Students identify the time on the half hour.
2. Students learn clock face, digital notation, and words for each time on the half hour.

Materials needed

Clock with a large face (i.e., schoolhouse clock)
Copy of Worksheet 2 for each student
Chalkboard and chalk
Pencil for each student

Introduce worksheet

1. Set the clock for 2 o'clock.
2. Show the clock to the students and ask what time it is.
3. Ask students to identify which is the minute hand and which is the hour hand.
4. Tell the students to watch closely while the clock is changed from 2 o'clock to 3 o'clock. (Change the time to 3 o'clock.)
5. What happened? (The minute hand went all the way around while the hour hand moved from the 2 to the 3.)
6. There are names for all the times in between the hours.
7. The clock now says 3 o'clock. Let's turn it some more, but have the minute hand go only halfway around. (Turn time to half past 3.)
8. Notice that the minute hand is on the 6. The hour is halfway between the 3 and the 4.
9. It is half past three. (Write the words on the chalkboard.)
10. It can also be written in digits—3:30. (Write this on the board.)
11. Turn clock to half past 4 and talk about "half past four" and "4:30."
12. Continue as long as interest indicates and it seems helpful.
13. Students complete the worksheet by writing the indicated time on each clock face in digits and in words.

Name _____   Date _____

Reproducible Worksheet 2

What time is shown on each clock face?
Write the time in digits and in words.

_____ :30   _____ :30   _____ :30   _____ :30
Half past ____   Half past ____   Half past ____   Half past ____

  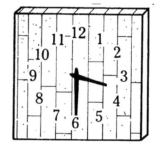 

_____ :30   _____ :30   _____ :30   _____ :30
Half past ____   Half past ____   Half past ____   Half past ____

 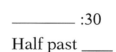  

_____ :30   _____ :30   _____ :30   _____ :30
Half past ____   Half past ____   Half past ____   Half past ____

© 1987, 1996 J. Weston Walch, Publisher   Clock and Calendar Skills

HOUR/HALF PAST
Directions for Worksheet 3

Objectives

1. Students distinguish between times on the hour and times on the half hour.
2. Students review clock face, digital notation, and words for time on the hour and half hour.

Materials needed

Large cardboard clock face (see page 2)
Copy of Worksheet 3 for each student
Chalkboard and chalk
Pencil for each student

Introduce worksheet

1. Set the clock for 7 o'clock.
2. Ask students what time it is. Have a student write the time on the chalkboard in words. Have someone else write the time in digits.
3. Set the clock for half past 4. Be sure the hour hand is between the 4 and the 5. Ask students what time it is. Have students write the time on the board in words and in digits.
4. Have students take turns setting the clock to times on the hour and on the half hour while other students identify the time and write it on the chalkboard in words and digits.
5. Students complete the worksheet by writing the indicated time on each clock face in digits and in words.

Alternative

If students have made their own clock faces, have them work in pairs—one setting a time while the other tells what time it says and writes the time in words and digits.

Name _____ Date _____

Reproducible Worksheet 3

What time is shown on each clock face?
Write the time in digits and in words.

 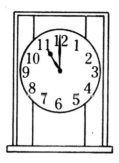  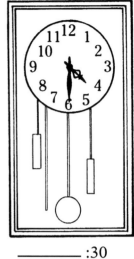

_____ :30   _____ :00   _____ :30   _____ :30

Half past ____   _____ o'clock   Half past ____   Half past ____

  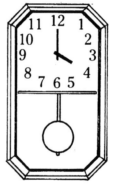 

_____ :30   _____ :00   _____ :00   _____ :00

Half past ____   _____ o'clock   _____ o'clock   _____ o'clock

_____ :30   _____ :00   _____ :30   _____ :30

Half past ____   _____ o'clock   Half past ____   Half past ____

© 1987, 1996 J. Weston Walch, Publisher   Clock and Calendar Skills

QUARTER PAST THE HOUR
Directions for Worksheet 4

Objectives

1. Students identify the time at quarter past the hour.
2. Students learn clock face, digital notation, and words for each time at quarter past the hour.

Materials needed

Clock with a large face (i.e., schoolhouse clock)
Copy of Worksheet 4 for each student
Chalkboard and chalk
Pencil for each student

Introduce worksheet

1. Set the clock for 1 o'clock. Ask the students what time it is.
2. Now let's set the clock to a different time. We're going to turn the minute hand only one quarter of the way around the clock face. (Change time to quarter past 1.)
3. This time is called quarter past one. (Write the words on the board.)
4. Or the time can be written in digits—1:15. (Write the time on the chalkboard.)
5. Turn the clock to quarter past 2. Ask the class what time the clock says. Have the students write the words (quarter past two) and the digits (2:15) on the chalkboard.
6. Continue turning the clock to other quarter-past-the-hour times. As students identify the time, have them write it on the chalkboard in words and in digits.
7. Students complete the worksheet by writing the indicated time on each clock face in digits and in words.

Name _____ Date _____

Reproducible Worksheet 4

What time is shown on each clock face?
Write the time in digits and in words.

_____ :15
Quarter past _____

_____ :15
Quarter past _____

_____ :15
Quarter past _____

_____ :15
Quarter past _____

_____ :15
Quarter past _____

_____ :15
Quarter past _____

_____ :15
Quarter past _____

_____ :15
Quarter past _____

_____ :15
Quarter past _____

_____ :15
Quarter past _____

_____ :15
Quarter past _____

_____ :15
Quarter past _____

QUARTER OF THE HOUR
Directions for Worksheet 5

Objectives

1. Students identify the time at quarter of the hour.
2. Students learn clock face, digital notation, and words for each time at quarter of the hour.

Materials needed

Clock with a large face (i.e., schoolhouse clock)
Copy of Worksheet 5 for each student
Chalkboard and chalk
Pencil for each student

Introduce worksheet

1. Set the clock for 4 o'clock.
2. Review with students what the time is and how to write the time in words and in digits (four o'clock; 4:00).
3. Turn clock to quarter past 4.
4. Review with students what the time is and how to write the time in words and in digits (quarter past four; 4:15).
5. Turn clock to half past 4.
6. Review with students what the time is and how to write the time in words and in digits (half past four; 4:30).
7. Continue turning the minute hand until it is on the 9.
8. What hour are we coming toward? (five o'clock)
9. When the minute hand is on the nine, we say that it is quarter of the hour. In this case, it is quarter of five. (Write the words *quarter of five* on the chalkboard.)
10. The digital way of writing this time is 4:45. (Write this on the chalkboard.)
11. Continue turning the minute hand, mentioning each hour, quarter past the hour, half past the hour, and quarter of the hour.
12. At each quarter of the hour, write the time on the chalkboard in words and in digits.
13. Students complete the worksheet by writing the indicated time on each clock face in digits and in words.

Name _____   Date _____

Reproducible Worksheet 5

What time is shown on each clock face?
Write the time in digits and in words.

_____ :45
Quarter of ____

_____ :45
Quarter of ____

_____ :45
Quarter of ____

_____ :45
Quarter of ____

_____ :45
Quarter of ____

_____ :45
Quarter of ____

_____ :45
Quarter of ____

_____ :45
Quarter of ____

_____ :45
Quarter of ____

_____ :45
Quarter of ____

_____ :45
Quarter of ____

© 1987, 1996 J. Weston Walch, Publisher   12   Clock and Calendar Skills

GAME: DRAW THE TIME

Objective

Students practice matching times in words with corresponding clock faces.

Materials needed

One copy of each of the next two pages (pages 14 and 15)
Scissors
Two boxes (shoe-box size)
Large cardboard clock face (see page 2)

To prepare the game

1. Make copies of the next two pages.
2. Cut apart the "hours" (one o'clock, two o'clock, etc.) that are listed on page 14. Put these 12 slips of paper in one box.
3. Cut apart page 15, which tells time on the hour, quarter past, half past, and quarter of. Put these slips of paper in the other box.

To play the game

1. Have one student draw a paper from each box and combine them to read the time.
2. Have another student arrange the hands on the clock to show the time that was read.
3. Continue in this manner.

Note

Pages 16, 17, and 18 contain additional times before and after the hour. As students become more familiar with these times, add them to the game.

Variations

1. Involve more students by having someone write the digits for the time on the chalkboard. Or have two students do the drawing, one picking the hour and one picking the time before or after the hour.
2. Instead of using the large clock face, draw a number of clock faces on the board for students to fill in.
3. Make several editions of the game and have students play in pairs.
4. Use the blank clock faces on page 60, 61, or 62 and have every student draw a clock face for each time that is drawn.
5. Playing pieces have clock faces on them showing where the minute hand should be pointing. If the students are more advanced, you may not want to include this help as part of the game. Use a piece of paper to cover the clock faces on the master before making your copies of the pages.

one o'clock	two o'clock
three o'clock	four o'clock
five o'clock	six o'clock
seven o'clock	eight o'clock
nine o'clock	ten o'clock
eleven o'clock	twelve o'clock

(clock)	Exactly	(clock)	Half past
(clock)	Quarter past	(clock)	Quarter of
(clock)	Exactly	(clock)	Half past
(clock)	Quarter past	(clock)	Quarter of
(clock)	Exactly	(clock)	Half past
(clock)	Quarter past	(clock)	Quarter of

(clock)	Five minutes past	(clock)	Ten minutes past
(clock)	Twenty minutes past	(clock)	Twenty-five minutes past
(clock)	Five minutes of	(clock)	Ten minutes of
(clock)	Twenty minutes of	(clock)	Twenty-five minutes of
(clock)	Five minutes past	(clock)	Ten minutes of
(clock)	Twenty minutes past	(clock)	Twenty-five minutes of

Clock and Calendar Skills

(clock)	One minute past	(clock)	Three minutes past
(clock)	Seven minutes past	(clock)	Eight minutes past
(clock)	Eleven minutes past	(clock)	Fourteen minutes past
(clock)	Seventeen minutes past	(clock)	Twenty-two minutes past
(clock)	Twenty-seven minutes past	(clock)	Eighteen minutes past
(clock)	Twenty-one minutes past	(clock)	Twenty-eight minutes past

(clock)	Two minutes of	(clock)	Four minutes of
(clock)	Six minutes of	(clock)	Nine minutes of
(clock)	Twelve minutes of	(clock)	Thirteen minutes of
(clock)	Sixteen minutes of	(clock)	Nineteen minutes of
(clock)	Twenty-three minutes of	(clock)	Twenty-four minutes of
(clock)	Twenty-six minutes of	(clock)	Twenty-nine minutes of

HOUR/QUARTER PAST/HALF PAST/ QUARTER OF
Directions for Worksheet 6

Objectives

1. Students distinguish between times on the hour, quarter past, half past, and quarter of the hour.
2. Students review clock face, digital notation, and words for time on the hour, quarter past, half past, and quarter of the hour.

Materials needed

Game "Draw the Time" (see page 13) Chalkboard and chalk
Large cardboard clock face (see page 2) Pencil for each student
Copy of Worksheet 6 for each student

Introduce worksheet

1. Before class, draw a series of clock faces similar to those below on the chalkboard.

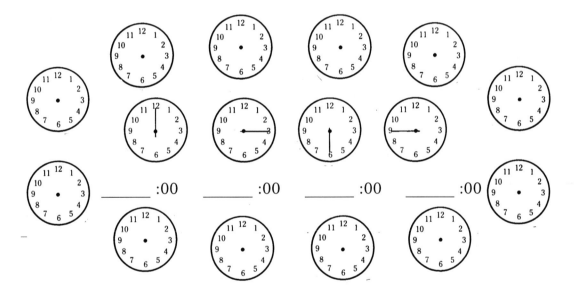

2. We've learned that the minute hand helps tell the time. Where is the minute hand pointing if it is half past the hour? (to the 6) Point to that clock in the center of the chalkboard.

3. Where does the minute hand point if it is exactly on the hour? (to the 12) Point to that clock in the center of the chalkboard.

4. Where does the minute hand point if it is quarter of the hour? (to the 9) Point to that clock in the center of the chalkboard.

5. Where does the minute hand point if it is quarter past the hour? (to the 3) Point to that clock in the center of the chalkboard.

6. Play the game "Draw the Time." For each time, have students show that time on one of the clock faces on the chalkboard.

7. If desired, have students write the digital name for each time under the clock face.

8. Students complete the worksheet by writing the indicated time on each clock face in digits and in words.

Reproducible Worksheet 6

What time is shown on each clock face?
Write the time in digits and in words.

_____
_____

_____
_____

_____
_____

_____
_____

_____
_____

_____
_____

_____
_____

_____
_____

_____
_____

_____
_____

_____
_____

_____
_____

Five, Ten, Twenty, and Twenty-five Past
Directions for Worksheet 7

Objectives

1. Students explore the relationship between hours and minutes.
2. Students identify the time at five, ten, twenty, and twenty-five past the hour.
3. Students learn clock face, digital notation, and words for time at five, ten, twenty, and twenty-five past the hour.

Materials needed

Large cardboard clock face (see page 3) Chalkboard and chalk
Copy of Worksheet 7 for each student Pencil for each student

Introduce worksheet

1. What do the numbers on this cardboard clock represent when the short hand points to one of them? (the hour)
2. There is another kind of mark on this clock that we haven't discussed. Can you see what it is? (the marks around the edge)
3. They are important for the longer hand. Does that give a hint what those marks are for? (They mark the minutes.)
4. There are 60 marks around the outside of the clock indicating the 60 minutes in an hour.
5. It is hard to count these marks every time, so there are certain landmarks that we learn.
6. Count the number of minute marks from 12 to 1. (There are 5.) So we know when the minute hand is on the 1, it is five minutes past the hour.
7. Arrange the clock hands so the hour hand is on the 10 and minute hand is on the 1. What time would this be? (five minutes past ten o'clock) In digits this would be 10:05. (Write it on the chalkboard.)
8. If we continue over to the 2 with the minute hand, how many minutes would that be? Count the marks. (ten minutes) So what time is it now? (ten minutes past ten o'clock) In digits this would be 10:10. (Write it on the chalkboard.)

9. If we continue moving the minute hand to the 3, that is a time that we've talked about before. (quarter past ten o'clock) Write that on the chalkboard in digits. What does that tell you about how many minutes it is past 10? (fifteen minutes)

10. How many minutes are indicated when the minute hand is pointing to the 4? (twenty minutes) What time is it if the hour hand is still on the 10 and the minute hand is pointing to the 4? (twenty past 10 o'clock; 10:20)

11. Let's continue moving the minute hand to the 5. What time is it now? (twenty-five minutes past ten o'clock; 10:25)

12. Continue practicing with other times five, ten, twenty, and twenty-five past the hour.

13. Students complete the worksheet by writing the indicated time on each clock face in digits and in words.

Reproducible Worksheet 7

What time is shown on each clock face?
Write the time in digits and in words.

_____ _____ _____ _____

_____ _____ _____ _____

_____ _____ _____ _____

_____ _____ _____ _____

_____ _____ _____ _____

_____ _____ _____ _____

Five, Ten, Twenty, and Twenty-five of
Directions for Worksheet 8

Objectives

1. Students identify the time at five, ten, twenty, and twenty-five of the hour.
2. Students learn clock face, digital notation, and words for time at five, ten, twenty, and twenty-five of the hour.

Materials needed

Clocks with a large face (i.e., schoolhouse clock)
Copy of Worksheet 8 for each student
Chalkboard and chalk
Pencil for each student

Introduce worksheet

1. Set the clock for 4 o'clock.
2. Review with students what the time is when the minute hand points to the 1, the 2, the 3, the 4, the 5. Use the word names and the digital names. Write both names on the chalkboard.
3. Move the minute hand to the 6 (the hour hand will be between the 4 and the 5). How do we write this time in digits? (4:30) Can you figure out how many minutes have passed since it was 4 o'clock? (thirty)
4. Continue moving the minute hand to the 7. How would we write this time in digits? (4:35)
5. When the minute hand moves past the 6, it is closer to the next hour than the last. It is closer to 5 o'clock than to 4 o'clock.
6. How many minute marks are there between the 7 and the 12? (twenty-five) So we say that it is twenty-five minutes before five o'clock (or twenty-five minutes of five).
7. Let's continue moving the minute hand until it is on the 8. How would this time be written in digits? (4:40) How would you say the time in words? (twenty minutes of five)
8. Continue moving the minute hand and discussing ten of and five of the hour.
9. Students complete the worksheet by writing the indicated time on each clock face in digits and in words.

Alternative

For some students it may be easier to start on the hour and make time go backward, discussing five minutes before the hour, followed by ten minutes before the hour, etc.

Reproducible Worksheet 8

What time is shown on each clock face?
Write the time in digits and in words.

_____ _____ _____ _____

_____ _____ _____ _____

_____ _____ _____ _____

GAME: TIME-O

Objective

Students practice matching word names for time with corresponding digital times.

Materials needed

One copy of each of the next three pages (pages 28, 29, and 30)
Scissors
Box for holding playing pieces
Time-O card for each student (see pages 31–42)
Pennies, beans, or paper scraps for covering cards

To prepare the game

1. Make a copy of each of the next three pages.
2. Cut apart the playing pieces and put them in the box.
3. Make a copy of a Time-O card for each student. The 24 cards starting on page 31 are all different. Copy a different card for each student.
4. Be sure each student has enough pennies, beans, paper scraps, or whatever to cover his or her squares.

To play the game

1. One person is the caller.
2. The caller draws a paper out of the box and reads aloud the time printed on it.
3. Players look on their cards to see if they have that time listed. If they do, they cover that clock with a penny.
4. Play continues until one player has covered five squares in a row across, vertically, or diagonally. The center square is a free square.
5. The player with five in a row calls out TIME-O.
6. The player calls out his or her five times to check against the playing pieces that have been called out by the caller.
7. Winner becomes the caller for the next game.

Variation

More capable students may want to play two or three cards at a time.

one o'clock	three o'clock
ten minutes past one	ten minutes past three
twenty minutes past one	twenty minutes past three
half past one	half past three
twenty minutes of two	twenty minutes of four
ten minutes of two	ten minutes of four
five minutes past two	five minutes past four
quarter past two	quarter past four
twenty-five minutes past two	twenty-five minutes past four
twenty-five minutes of three	twenty-five minutes of five
quarter of three	quarter of five
five minutes of three	five minutes of five

five o'clock	seven o'clock
ten minutes past five	ten minutes past seven
twenty minutes past five	twenty minutes past seven
half past five	half past seven
twenty minutes of six	twenty minutes of eight
ten minutes of six	ten minutes of eight
five minutes past six	five minutes past eight
quarter past six	quarter past eight
twenty-five minutes past six	twenty-five minutes past eight
twenty-five minutes of seven	twenty-five minutes of nine
quarter of seven	quarter of nine
five minutes of seven	five minutes of nine

nine o'clock	eleven o'clock
ten minutes past nine	ten minutes past eleven
twenty minutes past nine	twenty minutes past eleven
half past nine	half past eleven
twenty minutes of ten	twenty minutes of twelve
ten minutes of ten	ten minutes of twelve
five minutes past ten	five minutes past twelve
quarter past ten	quarter past twelve
twenty-five minutes past ten	twenty-five minutes past twelve
twenty-five minutes of eleven	twenty-five minutes of one
quarter of eleven	quarter of one
five minutes of eleven	five minutes of one

TIME-O Card #2

T	I	M	E	O
9:10	1:40	8:35	3:00	11:20
3:30	11:50	1:20	5:50	7:00
8:05	2:25	Free Space	10:15	4:35
5:10	7:40	6:55	6:15	7:30
8:45	12:25	2:45	4:05	12:55

TIME-O Card #1

T	I	M	E	O
3:10	9:30	4:15	2:05	10:55
3:50	8:05	7:20	4:55	10:15
9:10	3:40	Free Space	4:25	6:35
9:40	3:20	10:35	9:50	1:00
7:00	5:30	10:45	8:25	4:45

TIME-O Card #4

8:35	3:10	5:30	4:45	8:05
3:50	10:15	7:00	3:40	4:25
9:50	6:35	Free Space	1:00	9:20
9:40	7:30	3:20	10:25	4:15
10:55	2:05	4:55	10:45	9:10

TIME-O Card #3

2:05	5:30	4:45	10:15	9:30
11:40	6:35	9:50	11:00	3:10
9:10	4:55	Free Space	3:50	7:20
12:05	8:25	1:00	3:40	4:25
10:35	4:15	10:55	3:20	12:45

© 1987, 1996 J. Weston Walch, Publisher

Clock and Calendar Skills

TIME-O — Card #6

1:40	6:15	3:30	11:50	1:20
5:10	12:45	5:50	3:00	7:30
4:05	11:20	Free Space	6:55	9:10
10:15	12:25	11:40	10:05	2:45
12:55	8:35	2:25	9:00	4:35

TIME-O — Card #5

3:50	11:00	9:10	7:30	4:55
5:30	6:35	3:40	9:20	10:55
4:15	8:35	Free Space	4:45	3:20
1:00	2:05	4:10	10:25	11:40
12:45	10:15	12:05	3:10	9:50

© 1987, 1996 J. Weston Walch, Publisher — Clock and Calendar Skills

TIME-O Card #8

5:10	11:50	3:00	1:20	7:40
6:55	4:05	9:30	5:50	8:05
2:45	9:10	Free Space	10:35	7:20
8:45	6:15	12:55	3:30	8:25
4:35	10:15	2:25	7:00	1:40

TIME-O Card #7

12:05	7:10	4:25	11:30	3:10
4:15	1:30	5:00	9:20	7:40
6:05	9:50	Free Space	3:20	11:00
4:45	2:55	3:40	8:15	10:55
2:35	8:45	10:25	12:35	1:50

TIME-O Card #10

7:30	11:10	1:30	6:15	12:25
9:00	5:10	11:20	2:35	5:00
12:15	2:55	Free Space	6:05	1:20
8:35	9:40	11:50	10:05	1:40
12:55	2:45	2:25	1:50	10:45

TIME-O Card #9

6:05	10:25	3:10	8:15	3:20
4:25	5:00	9:20	1:30	10:45
3:40	7:10	Free Space	11:30	9:50
1:50	4:15	10:05	4:45	9:40
10:55	2:35	9:00	12:35	2:55

TIME-O Card #12

T	I	M	E	O
5:00	6:15	1:30	7:40	1:20
11:10	2:55	5:10	11:50	12:05
11:20	1:40	Free Space	7:30	12:15
8:45	11:00	2:25	6:05	2:35
12:25	8:35	1:50	12:55	2:45

TIME-O Card #11

T	I	M	E	O
1:10	3:00	6:25	9:30	5:40
3:30	8:55	7:20	2:15	10:35
8:15	11:40	Free Space	9:00	5:20
10:05	8:25	7:10	5:50	12:45
7:50	4:35	6:55	6:45	4:05

TIME-O

Card #14

12:05	12:15	5:10	7:30	1:50
6:15	9:20	1:30	5:00	2:25
8:35	11:50	Free Space	2:45	1:20
11:00	2:55	1:40	11:10	8:45
2:35	7:40	10:25	6:05	12:55

TIME-O

Card #13

1:10	7:50	4:05	8:25	7:10
3:30	6:55	7:20	9:30	3:00
5:50	8:05	Free Space	6:25	8:45
8:15	6:45	5:20	7:00	4:35
5:40	2:15	8:55	7:40	10:35

TIME-O Card #16

T	I	M	E	O
7:30	10:25	1:20	12:15	6:05
9:00	8:35	11:10	1:30	6:15
5:10	11:50	Free Space	2:55	9:20
9:40	2:45	10:05	1:40	2:35
5:00	2:25	1:50	10:45	2:35

TIME-O Card #15

T	I	M	E	O
2:05	5:20	10:45	4:55	6:45
9:30	11:10	3:50	2:15	5:30
6:25	1:10	Free Space	7:50	9:40
6:35	8:25	1:00	7:00	12:15
8:05	7:20	10:35	5:40	8:55

© 1987, 1996 J. Weston Walch, Publisher — Clock and Calendar Skills

TIME-O   Card #18

5:10	4:35	9:00	11:40	1:20
10:05	3:00	5:50	8:25	6:15
3:30	7:20	Free Space	10:35	1:40
2:45	12:55	12:45	11:50	6:55
2:25	9:30	10:15	4:05	9:10

TIME-O   Card #17

3:50	1:00	12:05	6:35	12:15
4:55	6:45	5:40	2:05	10:35
12:45	8:55	Free Space	8:25	5:20
9:30	11:40	7:20	1:10	2:15
11:10	6:25	5:30	11:00	7:50

© 1987, 1996 J. Weston Walch, Publisher — 39 — *Clock and Calendar Skills*

TIME-O   Card #20

8:05	1:00	10:45	8:55	6:25
12:15	6:35	11:30	6:45	9:20
9:40	12:35	Free Space	1:10	5:30
2:15	5:20	5:40	10:25	7:50
2:05	4:55	11:10	7:00	3:50

TIME-O   Card #19

11:00	12:15	9:20	5:40	7:50
12:35	3:50	2:05	6:25	11:30
5:20	11:40	Free Space	1:10	2:15
6:35	1:00	8:55	5:30	12:05
6:45	4:55	10:25	12:35	11:10

© 1987, 1996 J. Weston Walch, Publisher

TIME-O — Card #22

1:10	11:20	3:30	8:55	5:20
7:10	6:55	5:40	12:35	3:00
5:50	11:40	Free Space	4:05	12:45
6:25	9:00	10:05	6:45	11:30
4:35	2:15	7:50	12:25	8:15

TIME-O — Card #21

3:30	6:45	8:45	5:20	4:05
8:15	7:50	11:30	6:25	7:00
6:55	3:00	Free Space	11:20	12:35
4:35	7:40	1:10	2:15	7:10
5:50	7:05	12:25	5:40	8:55

© 1987, 1996 J. Weston Walch, Publisher   Clock and Calendar Skills

TIME-O Card #24

12:25	8:45	6:05	4:45	11:30
5:00	1:30	1:50	3:10	7:40
4:15	10:55	Free Space	7:10	2:35
2:55	3:20	11:00	9:50	8:15
12:05	4:25	11:20	12:35	3:40

TIME-O Card #23

4:25	1:50	10:55	5:15	5:00
9:00	4:45	6:05	1:30	7:10
8:15	2:35	Free Space	3:20	9:40
3:10	12:25	10:45	11:20	3:40
2:55	10:05	12:35	9:50	11:30

Five, Ten, Twenty, and Twenty-five Past and Of
Directions for Worksheet 9

Objectives

1. Students identify the time at five, ten, twenty, and twenty-five past and of the hour.
2. Students review clock face, digital notation, and words for time at five, ten, twenty, and twenty-five past and of the hour.

Materials needed

Clock with a large face (i.e., schoolhouse clock)
Copy of Worksheet 9 for each student
Chalkboard and chalk
Pencil for each student

Introduce worksheet

1. Set the clock for five minutes of 7 o'clock.
2. Ask students what time it is. (five minutes of seven)
3. Have someone write the time on the board in digits. (6:55)
4. Have another student set the clock to a different time that is five, ten, twenty, or twenty-five minutes past or of the hour.
5. Have a student identify the time on the clock.
6. Have a student write the time on the board in digits.
7. Continue in this manner until students are confident about identifying the time at five, ten, twenty, and twenty-five minutes past and of the hour.
8. Students complete the worksheet by writing the indicated time on each clock face in digits and in words.

Reproducible Worksheet 9

What time is shown on each clock face?
Write the time in digits and in words.

_____ _____ _____ _____ _____
_____ _____ _____ _____ _____

_____ _____ _____ _____ _____
_____ _____ _____ _____ _____

_____ _____ _____ _____ _____
_____ _____ _____ _____ _____

© 1987, 1996 J. Weston Walch, Publisher

Clock and Calendar Skills

MINUTES PAST AND OF THE HOUR
Directions for Worksheet 10

Objectives

1. Students identify the time at any time past or of the hour.
2. Students learn clock face, digital notation, and words for time at any minute past or of the hour.

Materials needed

Large cardboard clock face (see page 2)
Chalkboard and chalk
Clock on the wall (at correct time)
Copy of Worksheet 10 for each student
Pencil for each student

Introduce worksheet

1. Look at the large cardboard clock face we have been using. What have we talked about so far on this clock? (hour hand, minute hand, numbers)
2. What else on this clock should we discuss? (little marks between the numbers)
3. Does anyone remember what they are?
4. Let's look between the 12 and the 1. See that there is a space between the first mark and the second mark. How many of these spaces are there between the 12 and the 1? (five spaces) Does that give any hints about what the marks show?
5. If not, let's continue counting spaces to the 2. How many spaces are there between the 12 and the 2? (ten spaces)
6. We know that if the minute hand is on the 1, it is how many minutes past the hour? (five minutes) If the minute hand is on the 2, it is how many minutes past the hour? (ten minutes) So what do the marks show?
7. (Set clock at three minutes past 6 o'clock.) What hour is the hour hand pointing to? (six) How many minutes have passed since the hour? (three) What is the time in words? (three minutes past six) Write the time in digits. (6:03)
8. (Set clock at twelve minutes past 6 o'clock.) Now how many minutes is it past six? If the minute hand were on the 2, how many minutes would it be? (ten minutes) How many more minutes have passed since the minute hand was on the 2? (two minutes) Ten plus two equals? (twelve) So it is twelve minutes past six. Write the time in digits. (6:12)
9. Continue in this manner identifying various times past and of the hour.
10. Students complete the worksheet by writing the indicated time on each clock face in digits and in words.

Name _____   Date _____

Reproducible Worksheet 10

What time is shown on each clock face?
Write the time in digits and in words.

_____   _____   _____   _____

_____ _____ _____ _____

_____   _____   _____   _____

_____ _____ _____ _____

_____   _____   _____   _____

_____ _____ _____ _____

_____   _____   _____   _____

_____ _____ _____ _____

© 1987, 1996 J. Weston Walch, Publisher   *Clock and Calendar Skills*

DIGITAL TIME
Directions for Worksheet 11

Objective

Students identify digital times and their corresponding times on a clock face and in words.

Materials needed

Chalkboard and chalk
Copy of Worksheet 11 for each student
Pencil for each student

Introduce worksheet

1. Before class, draw eight to ten clock faces on the chalkboard. Include a point for the center, but do not draw hour or minute hands on any of the clocks.

2. Write the digital time "2:15" on the board.

3. How would we put this time on the clock face? Let's start with the minute hand—where will it be pointed? (toward the 3) Draw a minute hand on one of the clocks.

4. Where will the hour hand be pointing? (a little past the 2) Draw an hour hand on the clock.

5. How do we say this time in words? (quarter past two)

6. Write the digital time "4:50" on the board.

7. How would we put this time on the clock face? Let's start with the minute hand—where will it be pointed? (toward the 10) Draw a minute hand on another clock.

8. Where will the hour hand be pointing? (way past the 4; almost to the 5) Draw the hour hand on the clock.

9. How do we say this time in words? (ten minutes of five)

10. Continue writing other digital times on the chalkboard and having students draw the corresponding clock face and say the time in words. (*Note*: it may be easier to use minutes in multiples of 5—5 past, 10 past, etc., rather than have to deal with 12 past, 18 past, etc.)

11. Students complete the worksheet by showing the indicated digital time on the clock face and writing it in words.

Name _____   Date _____

Reproducible Worksheet 11

The time is given in digits.
Show the correct time on the clock face.
Write the time in words.

   8:20   2:35   5:05   12:30

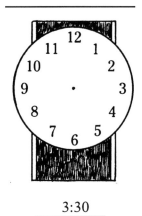 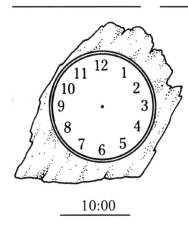  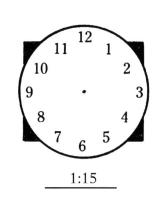

   3:30   10:00   7:10   1:15

   6:50   11:25   9:55   4:45

© 1987, 1996 J. Weston Walch, Publisher

Clock and Calendar Skills

DIGITAL CLOCKS
Directions for Worksheet 12

Objective

Students identify and read times on a digital clock.

Materials needed

Digital clock
Chalkboard and chalk
Copy of Worksheet 12 for each student
Pencil for each student

Introduce worksheet

1. Before class, set the clock to 6:00. Unplug the clock (or take out the batteries).

2. Show class members the digital clock, which now has a blank face (not plugged in or having batteries). What is this? (a clock) How is this clock different from the other clocks we have been using to tell time? (no hands) If there are no hands, how can we tell the time? (by the numbers; only the numbers necessary to tell the time show on the face of this clock)

3. Plug in the clock (or put in the batteries). What numbers show on the clock? (six, zero, zero) So what time is it according to this clock? (six o'clock)

4. During the time you have talked, the time has probably changed. What time does the clock say now?

5. Reset the time on the clock to twenty past six. What numbers show on the clock? (six, two, zero) What time is it? (six twenty)

6. Reset the time to 6:35. What numbers show now? (six, three, five) What time is it? (six thirty-five)

7. Draw a clock face on the board and put the hands to show 6:35. On a clock face, time past the half hour is usually called time "before the next hour." We might say that is is twenty-five before seven. But on a digital clock the time is usually said just as the numbers appear.

8. Continue resetting the clock and identifying the time.

9. Students complete the worksheet by writing the words for the time shown on each digital clock.

Name _____   Date _____

Reproducible Worksheet 12

What time is shown on each digital clock?
Write the time in words.

  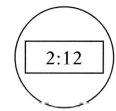 

_____ _____ _____ _____

_____ _____ _____ _____

_____ _____ _____ _____

_____ _____ _____ _____

_____ _____ _____ _____

Clock and Calendar Skills

TIME IN WORDS
Directions for Worksheet 13

Objective

Students put times written in words on clock faces and indicate the time in digits.

Materials needed

Game "Draw the Time" (see page 13)
Large cardboard clock face (see page 2)
Chalkboard and chalk
Copy of Worksheet 13 for each student
Pencil for each student

Introduce worksheet

1. Use all five pages of the game. Put the hours in one box and the times past and of the hour in another box.
2. Have one student draw a paper from each box and read the time aloud.
3. Have a second student arrange the hands of the cardboard clock face to show the corresponding time.
4. Have a third student write the time on the chalkboard using digits.
5. Let students take turns with each of the three tasks.
6. Continue until students can easily show clock face time and digital time corresponding to the words.
7. Students complete the worksheet by drawing the clock face and writing the digital time to correspond to each time written in words.

Name _____ Date _____

Reproducible Worksheet 13

The time is given in words.
Show the correct time on the clock face.
Write the time in digits.

   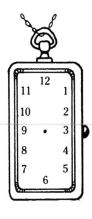

_____   _____   _____   _____
Twenty-five of one   Five past six   Quarter of three   Twenty-five past eleven

_____   _____   _____   _____
Ten past nine   Five of eight   Twenty past one   Ten o'clock

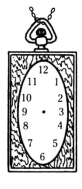   

_____   _____   _____   _____
Quarter past four   Half past eight   Twenty of four   Ten of five

© 1987, 1996 J. Weston Walch, Publisher   *Clock and Calendar Skills*

Card Game: Matching the Time

Objective

Students practice matching time in words with corresponding clock face and digital time.

Materials needed

Cards that are printed on the next four pages
Scissors

To prepare the game

1. Photocopy the four pages of cards on the sturdiest stock you have available.
2. Cut cards apart to make 40 cards.

To play the game

1. Two or three students can play.
2. Shuffle the cards.
3. Deal five cards to each player.
4. The first player draws a card, making six cards.
5. The player tries to make a book (three cards all telling the same time: clock face, digital, and words).
6. If able to make a book, the player lays the three cards down on the table and draws three more cards to replace them.
7. When a player is unable to make a book, he or she discards one card and play continues to the left.
8. The next players may pick up the discarded card or draw from the pile.
9. Play continues until all the cards in the pile have been drawn.
10. The player with the most books is the winner.

Variation

To make the game easier, place books face up on the table when any two of the corresponding cards are available. The third card in each book can be added as players discover them.

(clock showing 1:35)	1:35
twenty-five minutes of two	(clock showing 2:55)
2:55	five minutes of three
(clock showing 3:05)	3:05
five minutes past three	(clock showing 4:45)

4:45	quarter of five
(clock showing ~4:45)	5:25
twenty-five minutes past five	(clock showing ~5:25)
6:00	six o'clock
(clock showing ~7:40)	7:40

twenty minutes of eight	[clock showing 8:15]
8:15	quarter past eight
[clock showing 9:15]	9:15
quarter past nine	[clock showing 9:50]
9:50	ten minutes of ten

(clock showing ~10:30)	10:30
half past ten	(clock showing ~11:20)
11:20	twenty minutes past eleven
(clock showing ~12:10)	12:10
ten minutes past twelve	Wild Card

58   *Clock and Calendar Skills*

Teacher's Reference Sheet for "Matching the Time" Game

(clock)	1:35	twenty-five minutes of two	(clock)	2:55
(clock)	five minutes past three	3:05	(clock)	five minutes of three
4:45	quarter of five	(clock)	5:25	twenty-five minutes past five
7:40	(clock)	six o'clock	6:00	(clock)
twenty minutes of eight	(clock)	8:15	quarter past eight	(clock)
ten minutes of ten	9:50	(clock)	quarter past nine	9:15
(clock)	10:30	half past ten	(clock)	11:20
Wild Card	ten minutes past twelve	12:10	(clock)	twenty minutes past eleven

BLANK CLOCK FACES
Directions for
Worksheets 14, 15, 16

Objective

Students use extra clock faces as needed to obtain additional practice in telling time.

Worksheet 14

Variety of actual clock faces.

Worksheet 15

Each clock is numbered for easier reference to a specific clock.

Worksheet 16

Clocks include marks for indicating minutes.

Suggestions for use

1. Dictate times for students to record in words, in digits, and on the clock faces.

2. Use in playing the game "Draw the Time." See the game on page 13 for specific directions.

3. Draw in clock faces. Have students write in corresponding time in digits and in words.

4. Write in digital times. Have students draw clock faces to show the time and write the time in words.

5. Write times in words. Have students draw corresponding clock faces and write the digital times.

6. To simplify worksheets, eliminate either lines for words or blocks for digits before making copies.

7. Have students work together, taking turns filling in a clock face for the partner to write corresponding digital time and time in words.

Name _____ Date _____

Reproducible Worksheet 14

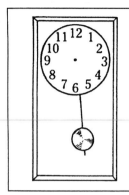   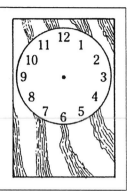

_____   _____   _____   _____

_____   _____   _____   _____

 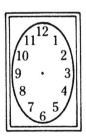  

_____   _____   _____   _____

© 1987, 1996 J. Weston Walch, Publisher   *Clock and Calendar Skills*

Name _____ Date _____

Reproducible Worksheet 15

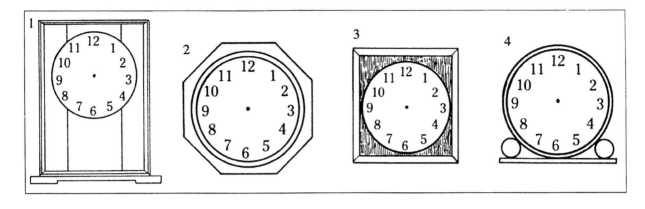

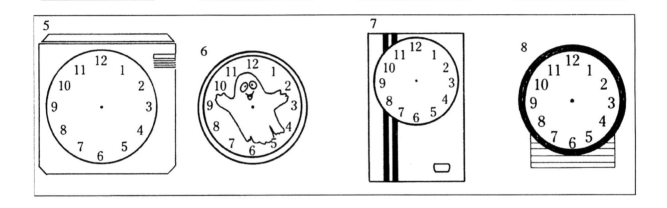

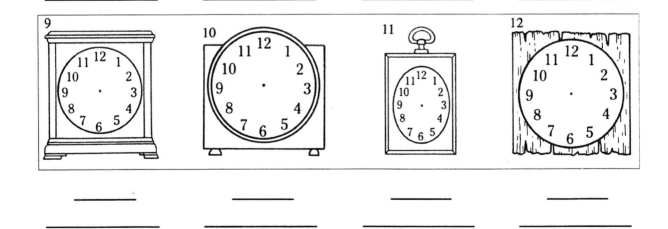

© 1987, 1996 J. Weston Walch, Publisher

Clock and Calendar Skills

Name _____ Date _____

Reproducible Worksheet 16

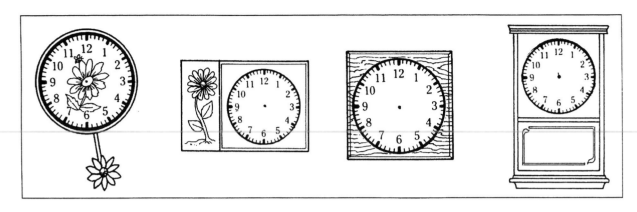

_____   _____   _____   _____

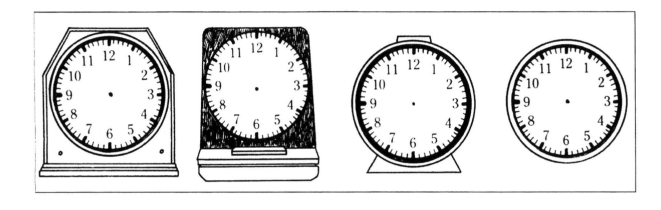

_____   _____   _____   _____

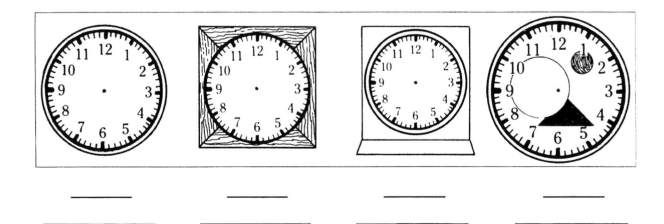

_____   _____   _____   _____

CLOCKS WITHOUT NUMBERS
Directions for Worksheet 17

Objectives

1. Students estimate the time by the relative positions of the hour and minute hands.
2. Students practice recognizing the time on clocks without numbers.

Materials needed

Large cardboard clock face (see page 2)
Copy of Worksheet 17 for each student
Pencil for each student

Introduce worksheet

1. Before class, remove the fastener and hands from the clock. Reattach the minute and hour hands so they are on the back of the cardboard disc where there are no numbers.
2. Show students the disc with the minute and hour hands. Ask them what is missing from this clock. (the numbers)
3. With all we've learned about telling time, perhaps we can do it even without the numbers.
4. (Set the hour hand facing straight down and the minute hand facing straight up.) If there were numbers on this clock, where would the minute hand be facing? (12) Where would the hour hand be facing? (6) So what time would it be? (six o'clock) (Try several other times on the hour: 9 o'clock, 3 o'clock, 1 o'clock, 4 o'clock, etc.)
5. (Set the minute hand facing straight down and the hour hand halfway between where the 3 and the 4 would be.) Where is the minute hand facing? (toward the six) So what do we know about the time? (It's half past the hour.) What hour is it half past? (three)
6. Continue practicing, setting the time to be quarter past and of the hour.
7. As students gain skill, try five, ten, twenty, and twenty-five past and of the hour.
8. Students complete the worksheet by writing the indicated time in digits and in words on each clock face.

Name _____   Date _____

Reproducible Worksheet 17

What time is shown on each clock face?
Write the time in digits and in words.

_____   _____   _____
_____   _____   _____

 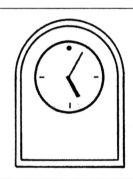 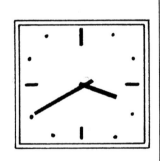

_____   _____   _____
_____   _____   _____

 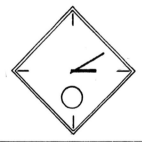 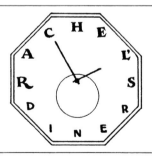

_____   _____   _____
_____   _____   _____

© 1987, 1996 J. Weston Walch, Publisher   *Clock and Calendar Skills*

SETTING AN ALARM
Directions for Worksheet 18

Objectives

1. Students distinguish between the alarm setting and the minute and hour hands.
2. Students set the alarm for different times.
3. Students identify a variety of alarm settings.

Materials needed

Large cardboard clock face (see page 2)   Alarm clock
Copy of Worksheet 18 for each student   Pencil for each student
Piece of cardboard (red or yellow) cut to represent alarm setting

Introduce worksheet

1. Before class, remove the hour and minute hands from the cardboard clock. Attach the alarm hand.
2. Today we are going to talk about a special part of some clocks—the alarm setting.
3. (Set alarm on the 2.) If this were the hour hand, what time would it be pointing to? (two o'clock) The alarm setting tells time in the same way as the hour hand. This setting is for two o'clock.
4. (Move alarm setting to the 6.) Now what time is the alarm set for? (six o'clock)
5. Let's try something a little different. (Move alarm to halfway between the 3 and the 4.) What time is the alarm pointing to now? (Halfway between three o'clock and four o'clock is half past three.)
6. Continue with several other settings on the cardboard clock face.
7. (Set the alarm clock for 3 o'clock and set the alarm setting for 6 o'clock.) There is a big difference between the hour hand and alarm setting. Watch the hands on this clock. (Move time toward 4 o'clock and then 5 o'clock.) What happens to the hour hand as the time changes? (It moves.) What happens to the alarm setting? (It stays in the same place.)
8. (Pull button to set alarm.) What time is the alarm set for? (six o'clock) It is now five o'clock. Watch what happens as time continues. (Change time to 6 o'clock, activating alarm.)
9. Set the clock for various times. Have students take turns setting the alarm at specific times and then moving the time to activate the alarm. (Go to bed at _____; set alarm for _____.)
10. Students complete the worksheet by drawing lines to connect each clock with the indicated time and alarm settings.

Reproducible Worksheet 18

Draw lines to connect the times with the correct clock.

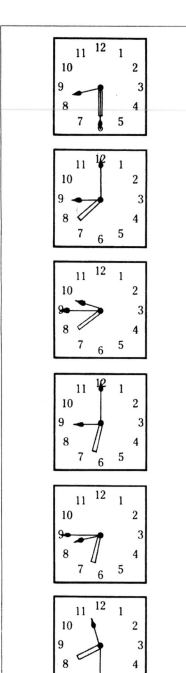

Go to bed at 10:30; set alarm for 8:00.

Go to bed at 9:00; set alarm for 7:30.

Go to bed at 9:15; set alarm for 6:30.

Go to bed at 11:00; set alarm for 7:00.

Go to bed at 8:45; set alarm for 6:30.

Go to bed at 8:30; set alarm for 6:00.

Go to bed at 11:00; set alarm for 8:30.

Go to bed at 9:00; set alarm for 6:30.

Go to bed at 9:30; set alarm for 6:30.

Go to bed at 10:30; set alarm for 7:00.

Go to bed at 11:30; set alarm for 8:00.

Go to bed at 9:45; set alarm for 7:30.

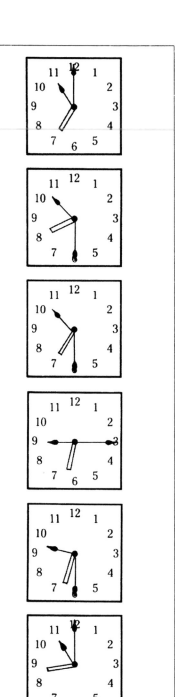

© 1987, 1996 J. Weston Walch, Publisher

Clock and Calendar Skills

USING A TIMER
Directions for Worksheet 19

Objectives

1. Students distinguish between a clock face and a timer.
2. Students set a timer for various times.
3. Students identify times when a timer would be helpful.

Materials needed

Chalkboard, chalk, and an eraser
Timer
Copy of Worksheet 19 for each student
Pencil for each student

Introduce worksheet

1. Before class, draw a large timer on the chalkboard. (See diagram below.)

2. What does this circle on the board look like? (It looks a little like a clock face.) What is different from a clock face? (It only has one hand; the numbers are not the same.)

3. This is a timer. Do you have a timer at your house—perhaps in the kitchen? (Show actual timer.) What are some ways that a timer is used? (For cooking, for timing exercise or practice, as a reminder to do something in half an hour, etc.)

4. The timer is used for timing short periods of time. What do you think the numbers stand for? (minutes)

5. Look at the timer on the board. How do you suppose we can tell what it is set for? (by looking at the arrow) Where is the arrow pointing? (to the 20) So how many minutes is the timer set for? (20 minutes) If this were a real timer, how would we know when 20 minutes has passed? (Bell would go off.)

6. Have someone in the class set the real timer for 20 minutes. Set it aside while class continues.

7. Let's go back to the timer on the board. (Erase arrow.) Would someone show how to set the timer for 35 minutes? (Have student draw arrow pointing to 35.)

8. What if you wanted to remember to call a friend in 45 minutes? Set the timer for that.//
9. Continue with several other settings on the timer, letting students take turns drawing the arrow at the proper setting.
10. Check the real timer. How much time is left of the original 20 minutes? Ask students to figure out how much time has passed.
11. Students complete the worksheet by drawing arrows on each timer to the indicated setting.

Note

The dials on some timers are on the center that moves. The numbers on those timers go in the opposite direction. While working on the same principle, they might be confusing to use with this worksheet.

Name _____   Date _____

Reproducible Worksheet 19

Draw an arrow to set each timer for the needed time.

 Bake the cookies 10 minutes.

 Boil the corn 5 minutes.

 Exercise for half an hour.

 Practice piano 20 minutes.

 Bake bread 50 minutes.

 Do homework 35 minutes.

 Nap for 45 minutes.

 Bake the potatoes 55 minutes.

 Bake the cake 40 minutes.

 Read a book 15 minutes.

 Talk on the phone 10 minutes.

 Bake muffins 25 minutes.

© 1987, 1996 J. Weston Walch, Publisher   Clock and Calendar Skills

A.M./P.M.
Directions for Worksheet 20

Objectives

1. Students distinguish between A.M. events and P.M. events.
2. Students list activities they do in the A.M.
3. Students list activities they do in the P.M.

Materials needed

Chalkboard and chalk
Copy of Worksheet 20 for each student
Pencil for each student

Introduce worksheet

1. Write the time "7:00" on the chalkboard. Ask students what they might be doing at seven o'clock. (eating breakfast, catching bus to school, watching TV, washing dinner dishes. Depending on how students respond, introduce possibilities at the other end of the day.)

2. Write the time "12:00" on the chalkboard. How about twelve o'clock—what would you be doing then? (eating lunch *or* sleeping)

3. It's hard to tell which seven o'clock or twelve o'clock we're talking about. There are two in every day. In one day, the hour hand goes twice all the way around the clock.

4. Write "A.M." after the "7:00" on the chalkboard. Does this clear up the confusion? When is 7:00 A.M.? (morning) What would we write to talk about seven o'clock in the evening? (7:00 P.M.)

5. Between twelve noon and twelve midnight is P.M. Between twelve midnight and twelve noon is A.M.

6. Is it A.M. or P.M. when you go to bed? (P.M.) When you eat breakfast? (A.M.) When the sun sets? (P.M.)

7. Talk about other activities that take place in the A.M. and P.M. Be sensitive to variations: People who work evening shifts go to work in the P.M. and come home in the A.M., etc.

8. Students complete the worksheet by determining which activities are done in the A.M. and which activities are done in the P.M.

Name _____   Date _____

Reproducible Worksheet 20

Write "A.M." beside things you do in the morning.
Write "P.M." beside things you do in the afternoon and evening.
Some things may be done in both the A.M. and the P.M.
Put an X beside things you do not do at all.

_____	brush teeth	_____	wake up
_____	have bedtime snack	_____	watch sun rise
_____	take shower or bath	_____	eat breakfast
_____	watch sun set	_____	go to bed
_____	set alarm	_____	eat dinner
_____	go to work	_____	read newspaper
_____	feed pets	_____	do homework
_____	get home from school	_____	eat lunch
_____	hear alarm clock ring	_____	make bed
_____	read newspaper	_____	get mail
_____	go to school	_____	get home from work
_____	watch TV	_____	pack lunch for school

List two other activities you do in the A.M.

A.M. _____

A.M. _____

List two other activities you do in the P.M.

P.M. _____

P.M. _____

© 1987, 1996 J. Weston Walch, Publisher   *Clock and Calendar Skills*

A.M./P.M. on Digital Clocks
Directions for Worksheet 21

Objectives

1. Students read the time on digital clocks.
2. Students distinguish between A.M. and P.M. on a digital clock.

Materials needed

Digital clock with a light that shows A.M. or P.M.
Chalkboard and chalk
Copy of Worksheet 21 for each student
Pencil for each student

Introduce worksheet

1. Before class, draw eight to ten digital clocks on the chalkboard. Put a small circle beside each set of digits. Fill in the circle to indicate it is P.M. on half the clocks. Leave the rest of the circles empty (lights off) to show A.M.

2. Before class, set the digital clock for 11:45 A.M.

3. Plug in the clock (or put in the batteries) and show it to the class. What numbers show on the clock? (one, one, four, five) What time is it? (eleven forty-five)

4. Turn the clock time to 12:10. What is different on this clock now? (different numbers) What else? (there's a light on) The light shows it is P.M.—between noon and midnight. *Note:* be sure your clock operates the way these instructions indicate. If it works differently (i.e., light on indicates A.M. or some other variation), change these instructions. When light is not on, it is A.M.—between midnight and noon.

5. Look at the clocks drawn on the chalkboard. What time does each say? The light determines whether it is A.M. or P.M.

6. Students complete the worksheet by deciding which activity they are more likely to be doing at the time shown on each clock.

Name _____   Date _____

Reproducible Worksheet 21

What time is shown on each digital clock?
The empty circle ○ shows the P.M. light is off; the filled circle ● shows the P.M. light is on.
Choose the activity you are more likely to be doing at the time shown.

| 7:00 ○ | 5:00 ● | 11:30 ● |

☐ hear alarm clock ☐ prepare dinner ☐ eat lunch
☐ watch sun set   ☐ wake up to alarm ☐ watch *Late Show*

| 7:20 ○ | 10:00 ● | 12:30 ● |

☐ watch TV   ☐ set alarm clock ☐ eat lunch
☐ eat breakfast   ☐ attend class   ☐ sleep

| 2:30 ● | 4:00 ○ | 7:30 ● |

☐ sleep   ☐ go to after-school job ☐ watch TV
☐ get home from school   ☐ be sound asleep   ☐ eat breakfast

| 7:15 ○ | 8:00 ○ | 11:30 ● |

☐ make bed   ☐ go to a movie   ☐ wash lunch dishes
☐ watch evening news   ☐ get dressed   ☐ wish on a star

© 1987, 1996 J. Weston Walch, Publisher — *Clock and Calendar Skills*

EARLIER/LATER
Directions for Worksheet 22

Objectives
1. Students identify earlier and later times.
2. Students add and subtract to determine earlier and later times.

Materials needed
Chalkboard and chalk
Clock with a large face (i.e., schoolhouse clock)
Copy of Worksheet 22 for each student
Pencil for each student

Introduce worksheet
1. Before class, write the following pairs of times on the chalkboard:

| 2:15 A.M. | 6:02 A.M. | 7:55 A.M. | 4:00 P.M. | 4:55 P.M. | 7:15 P.M. | 2:18 A.M. | 6:00 A.M. |
| 5:20 P.M. | 4:50 A.M. | 2:20 P.M. | 12:30 P.M. | 4:59 P.M. | 8:23 A.M. | 1:30 P.M. | 12:15 A.M. |

2. Today we are going to talk about two words: earlier and later. (Write the words on the chalkboard.)
3. Let's look at the first two times (2:15 A.M. and 5:20 P.M.). Which one is earlier in the day? (2:15 A.M.) How can you tell? (A.M. is in morning and P.M. is after noon.) Circle 2:15 A.M.
4. The second pair are both in the A.M. Which one is earlier? (4:50 A.M.)
5. In the next pair, which is earlier? (7:55 A.M.) You might think 2:20 would be earlier than 7:55; but 7:55 A.M. would be earlier than 2:20 P.M.
6. The fourth pair (4:00 P.M. and 12:30 P.M.) is a little extra tricky. They are both after noon, but which is earlier? Which comes before the other after noon? (12:30 P.M.)
7. Look at the next four pairs of times and determine which is later.
8. (Set clock for quarter past eight.) What time is it on this clock? (quarter past eight) What time will it be ten minutes later? To find out, let's just move the minute hand ten minutes. (It's twenty-five past eight.)
9. (Set clock for ten minutes of nine.) What time is on this clock? (ten minutes of nine) What time will it be one half hour later? To find out, let's move the minute hand one half hour. (It's twenty minutes past nine.)
10. Continue adding time in this manner until students are able to do it easily.
11. (Set clock for half past ten.) What time is it on this clock? (half past ten) The question is a little different this time: What time was it fifteen minutes earlier? To find out, move the minute hand back fifteen minutes. (It's quarter past ten.)
12. Continue setting the clock and subtracting time to find earlier times.
13. Students complete the worksheet by finding earlier and later times.

Name _____   Date _____

Reproducible Worksheet 22

1. Circle the later time in each pair.

   | 1:15 A.M. | 4:55 P.M. | 2:05 A.M. | 3:25 A.M. |
   | 8:00 A.M. | 2:15 P.M. | 3:00 P.M. | 1:30 P.M. |

2. What time will it be one half hour later than:

   5:00 P.M. _____   2:15 A.M. _____   6:45 A.M. _____
   2:20 P.M. _____   3:30 P.M. _____   11:45 A.M. _____

3. Circle the earlier time in each pair.

   | 2:00 A.M. | 3:45 P.M. | 11:15 P.M. | 12:35 P.M. |
   | 1:00 P.M. | 5:34 P.M. | 1:20 A.M. | 12:50 A.M. |

4. What time will it be 15 minutes earlier than:

   2:25 P.M. _____   6:00 A.M. _____   3:15 A.M. _____
   4:05 A.M. _____   12:10 P.M. _____ 6:45 P.M. _____

5. The basketball game starts at 11:00 A.M. The coach asks you to come to the gym one half hour earlier. What time should you get to the gym?

   _____

6. School starts at 8:05 A.M. You arrive 15 minutes later. What time do you arrive?

7. Dinner will be at 5:30 P.M. You need to put the casserole in the oven 45 minutes earlier. When should you put the casserole in the oven?

   _____

8. The party is at 6:00 P.M. You have been asked to come one hour earlier to help with decorations. What time should you arrive?

   _____

9. You usually get through work at 4:30 P.M. The boss asks you work 45 minutes later. What time will you get through?

   _____

DAILY SCHEDULE
Directions for Worksheet 23

Objectives

1. Students relate activities in their daily lives with specific times.
2. Students distinguish between A.M. and P.M.
3. Students order time from earliest to latest.

Materials needed

    Chalkboard and chalk
    Copy of Worksheet 23 for each student
    Pencil for each student

Introduce worksheet

1. Before class, draw the schedule outline shown below on the chalkboard.

TIME	ACTIVITY

2. What was the first thing you did this morning? (woke up) Write "Wake up" on the first line under ACTIVITY. What time did you do that? (Choose answer of one student—seven o'clock.) Seven P.M.? (No, seven A.M.; write "7:00 A.M." under TIME on the first line.)

3. What was the next thing you did this morning? Write the activity and the time it was done.

4. Continue through three or four morning activities, writing the activity and the time each activity was done.

5. Students complete the worksheet by making their own list of the day's activities and the time each activity is done.

Name _____   Date _____

Reproducible Worksheet 23

Keep track of your schedule for one day. Write the time as you begin each activity.

TIME	ACTIVITY
A.M.	Wake up
P.M.	Go to bed

TV SCHEDULE
Directions for
Worksheets 24, 25, 26

Objectives

1. Students consult a printed schedule to answer questions about TV programming.
2. Students use a variety of schedule formats.
3. Students add and subtract times to determine length of programs.
4. Students plan their own TV viewing schedule.

Worksheet 24

Uses grid style of scheduling: channels plotted against times.

Worksheet 25

Lists programs and information about them under starting time.

Worksheet 26

Incorporates TV schedule from student's own local paper.

Materials needed

Copy of appropriate worksheet for each student
Copy of tonight's local TV schedule for each student (if using Worksheet 26)
Pencil for each student

Introduce worksheet

1. How do you decide what to watch on TV? (Flick channels.) What if you want to know when a certain program will be on TV? How can you find out? (Look at a TV schedule.)
2. (Pass out worksheet.) TV schedules come in a variety of formats. Today we are going to learn to read this kind of schedule.
3. Talk with class about information shown on schedule. Ask specific questions to be sure they understand how to use the schedule.
4. Students will complete the worksheet by consulting the TV schedule to answer the questions.

Name _____ Date _____

Reproducible Worksheet 24

Use this TV schedule to help you answer the questions below.

Wednesday prime time

	6:00	6:30	7:00	7:30	8:00	8:30	9:00	9:30	10:00	10:30	11:00	11:30
2	News	NBC News	Wheel of Fortune	Jeopardy!	Healthviews		Dateline		Law & Order		News	Tonight Show
4	News		CBS News	Entertainment Tonight	Rescue 911	The Langoliers (1995) (Patricia Wettig, Bronson Pinchot)			National—Journal		News	Late Show
5	News		ABC News	Inside Edition	Chronicle	Roseanne	Ellen	Grace Under Fire	Coach	Primetime Live		News	Nightline
6	News		NBC News	Murphy Brown	Roseanne	Healthviews		Dateline		Law & Order		News	Tonight Show
7	News		NBC News	Wheel of Fortune	Jeopardy!	Mad About You		Dateline		Law & Order		News	Nightline
8	News		ABC News	Wheel of Fortune	Jeopardy!	Roseanne	Ellen	Grace Under Fire	Coach	Closeup At a Loss for Words: Illiterate		News	Nightline
9	News		ABC News	Entertainment Tonight	Hard Copy	Roseanne	Ellen	Grace Under Fire	Coach	Primetime Live		News	Nightline
10	Time Goes	Business Report	MacNeil/Lehrer NewsHour		Frontline		National Geographic				Motorweek	First Flights
11	MacNeil/Lehrer NewsHour		Business Report	One on One	Frontline		National Geographic				Currents	Think Tank
12 26	British Isles	Business Report	Growing Old in a New Age		Explorers		Outside Line		MacNeil/Lehrer NewsHour		Acorn—Nature	Waterways
13	News		CBS News	Entertainment Tonight	Inside Edition	Rescue 911	A River Runs Through It (1992) (Craig Sheffer, Brad Pitt)				News	Tennis Highlights
38	Beverly Hills, 90210		Cheers	H. Patrol	Legend		Marker		News	Cheers	M*A*S*H	Hogan's Heroes
56	Full House	Fresh Prince	Star Trek: The Next Generation		In the Heat of the Night (1995) (Carroll O'Connor)				News		Star Trek	

1. At what times could you watch the news on channel 6? _____

2. On what stations could you watch the *Tonight Show*? _____

3. How many different programs begin at 9 o'clock? _____

4. On what stations and at what times could you watch *M*A*S*H*?
_____

5. Which stations do not have news reports? _____

6. What movies are on tonight? What time does each begin and on what channel?

© 1987, 1996 J. Weston Walch, Publisher   79   *Clock and Calendar Skills*

Name _____   Date _____

Reproducible Worksheet 25

Use the TV schedule below to help you answer the questions below.

Saturday

12:00 Noon

(2) Hometime
(4) Beakman's World (CC)
(5) News (CC)
(6) Weekend Travel Update
(7) Saved by the Bell: The New Class(CC)
(10) NBA Inside Stuff
(9) Sports Page
(11) European Journal
(12) Cro (CC)
(25) When I Was a Girl
(27) Dios con Nosotros
(38) Movie: ★★★ Smokey and the Bandit" (1977, Comedy)
(50) Candlepin Bowling
(56) Movie: ★★★ "Miss Firecracker" (1989, Comedy-Drama)
(62) Saturday Specials
(68) Movie: ★★ "At War With the Army" (1950, Comedy)
(A&E) 20th Century
(COM) Young Ones
(DSC) America Coast to Coast
(E!) News Week in Review
(ESPN) LPGA Golf
(FAM) Centennial
(MC) (12:25) Movie: ★★★ "Conrack" (1974, Drama) (In Stereo) (PG)
(NECN) Weekend News
(NESN) Red Sox Weekly
(SHO) Movie: "Perry Mason: The Case of the Skin-Deep Scandal" (1993, Mystery) (In Stereo)
(WWOR) Kojak

12:30 p.m.

(2) Healthy Indulgences (CC)
(4) Storybreak (OC)
(5) Paid Program
(7) Name Your Adventure (CC)
(10) Siskel & Ebert (CC)
(11) Victory Garden (CC)
(12) Weekend Special (CC)
(44) Wild America (CC)
(AMC) Batman and Robin
(COM) Vacant Lot
(NESN) Red Sox Digest

1:00 p.m.

(2) Auction
(4) American Gladiators (OC)
(5) Fishing: Outdoor Adventures
(7) Saved by the Bell: The New Class (CC)
(10) Movie: ★★ "Penalty Phase" (1986, Drama)
(11) Frugal Gourmet(CC)
(12) Nick News
(25) World Wrestling Federation Superstars
(27) Pachanga Latina
(44) Ciao Italia
(A&E) Investigative Reports
(AMC) Movie: ★★ "It Happens Every Spring" (1949, Comedy)
(COM) Absolutely Fabulous
(DIS) (1:05) Danger Bay (CC)
(DSC) Secret Weapons
(E!) Features
(HBO) Lifestories: Families in Crisis (MAX) (1:15) Movie: ★★ "Clue" (1985, Comedy) (PG-Mild violence)
(NECN) Weekend News
(NESN) Seattle Mariners at Boston Red Sox
(SC) Italian League Soccer
(WWOR) Barnaby Jones

1:30 p.m.

(5) Extremists
(7) Casting About
(9) Night Court
(11) Cooking in France (CC)
(12) Nova Scotia Tourism
(44) Cucina Amore
(COM) To Be Announced
(DIS) Zorro (CC)
(DSC) Firepower
(E!) Coming Attractions
(HBO) Movie: "Showdown" (1993, Drama) (In Stereo) (PG-13-Adult language, violence) (CC)
(SHO) (1:45) Movie: ★★★ "Micki & Maude" (1984, Comedy) (PG 13-Adult situation)

2:00 p.m.

(4) Fisherman's Paradise
(5) Star Search
(7) Weekend Special
(9) Track and Field
(11) Ciao Italia
(12) Dear John (CC)
(25) World Wrestling Federation Wrestling Challenge
(27) Onda Max
(38) Movie: ★★★ "Racing With the Moon" (1984, Drama)
(44) Cooking at the Academy
(56) Movie ★★★ "Biloxi Blues" (1988, Comedy-Drama)
(68) Movie: ★★ "Visit to a Small Planet" (1960, Comedy)
(A&E) American Justice
(COM) A-List (CC)
(DIS) Swamp Fox (CC)
(DSC) Challenge
(E!) Guide to Summer Movies
(ESPN) Golf
(MC) (2:15) Movie: ★★ "Across the Tracks" (1991, Drama) (PG-13-Adult language, adult situations)
(NECN) Weekend News
(SC) Rugby World Cup—Canada vs. South Africa
(WWOR) Streets of San Francisco

1. What programs might be of interest to people who like to fish? _____
   _____

2. When and on what channels could you watch the news? _____
   _____

3. What is the choice of movies? What time does each begin and on what channel is it shown? _____
   _____

4. What programs would be of interest to someone who enjoys cooking?
   _____

5. What different kinds of sports are on TV? _____
   _____

© 1987, 1996 J. Weston Walch, Publisher   80   *Clock and Calendar Skills*

Name _____ Date _____

Reproducible Worksheet 26

Look at the newspaper for today's TV schedule. Use it to answer the questions below.

1. What different channels can you watch on your TV?
   _____

2. Which program would you most like to watch tonight?
   _____

   What time is it on? _____

   Which channel is it on? _____

   What time does the program end? _____

   How long is the program? _____

3. What other programs would you like to watch tonight? _____
   _____

4. Make a schedule of the TV programs you would like to watch this evening. Write the time of the earliest program, the name of the program, and the channel it is on. Then do the same for the next program. Continue for all the programs you want to watch.

Time	Name of the Program	Channel

© 1987, 1996 J. Weston Walch, Publisher

Clock and Calendar Skills

BUS SCHEDULE
Directions for Worksheet 27

Objectives
1. Students consult a bus schedule to compare specific trips.
2. Students use a bus schedule to answer travel questions.
3. Students add and subtract times to determine length of trips.

Materials needed
    Copy of Worksheet 27 for each student
    Pencil for each student

Introduce worksheet
1. If you wanted to go on the bus, how would you know when to catch the bus and which bus to take? Could you just stand on the corner, get on the next bus to go by, and hope it would take you where you want to go? (No, you would use a bus schedule.)
2. (Pass out worksheets.) Today we're going to look at this bus schedule between Portland, Maine, and New York City.
3. Where are the cities listed? (down the center of the schedule)
4. If you want to go from Portland to New York (toward the south), which schedule do you look at? (the numbers on the left)
5. If you want to go from New York (toward the north), which schedule do you look at? (the numbers on the right)
6. How can you tell if a bus leaves in the morning or afternoon? (Lightfaced times are A.M.; boldfaced times are P.M.)
7. What time does bus #613 leave Portland and head south? (9:55 P.M.) What is its first stop? (Portsmouth, N.H.) What time does it stop there? (10:50 P.M.) How long does it take to get from Portland to Portsmouth? (55 minutes)
8. What is the first bus in the afternoon to leave Boston heading north? (#726 leaves at 12:45 P.M.) How many stops does it make on its way to Portland? (three stops between Boston and Portland)
9. Continue asking questions about the schedule until students are comfortable using it.
10. Students complete the worksheet by consulting the bus schedule to answer the questions.

Suggestions
11. Make copies of the local bus schedule available to students. Ask specific questions to be answered by consulting the schedule. When is the next bus to the shopping plaza? What time is the last bus home tonight? etc.
12. If appropriate, take students on a bus trip that they plan from the bus schedule.
13. Provide copies of train or airline schedules. Compare travel times to different cities.

Name _____   Date _____

Reproducible Worksheet 27

Use this bus schedule to help you answer the questions below.

The lightfaced times are A.M.; the boldfaced times are P.M.

The schedule to the left is southbound (Portland to New York); read it top to bottom.

The schedule to the right is northbound (New York to Portland); read it bottom to top.

SOUTHBOUND READ DOWN								SCHEDULE NUMBER								NORTHBOUND READ UP
613	539	637	525	635	535	713	813	Folder No. **2002** 9-3-86 FREQUENCY	524	650	726	552	636	738	848	748	512
							X7								X6		
...	9 55	5 15	1 10	12 01	10 00	10 00	5 55	Lv ▲PORTLAND	≡10 55	≡12 35	3 20	≡5 05	...	7 05	...	8 50	11 55
		5 25	1 20	12 10	↓	10 10	↓	Portland Airport	10 45	12 25	3 10	4 55		↑	D 6 45	8 30	11 40
						10 30		Lv ▲Saco							D 6 42	8 27	
						10 33		▲Biddeford, ME									
...	10 50	6 25	2 20	1 10	Express	11 20	7 05	▲Portsmouth, NH	9 55	11 20	2 10	3 55	...	5 55	7 25	7 40	10 40
	D	7 00	2 55	1 45	↓	11 55	7 40	▲Newburyport, MA	9 20	10 50	1 35	3 20		5 20	6 50	7 05	10 05
	↓	↓	↓	↓			D 8 20	Haymarket Square						↑	↑	↑	↑
...	12 10	7 50	3 45	2 35	12 05	12 45	8 30 D 7 45 7 55 Ar	▲Boston, MA	Lv	8 30	10 00	12 45	2 30	4 30	6 00	6 15	
		7 50	4 30	2 35	12 45	12 45	8 30	Lv ▲Boston, MA	Ar	...	9 30	12 15	2 15	4 00	5 30	5 30	
		D 8 20	5 10	D 3 05	1 15	1 15	9 00	Ar Logan Airport	Lv	...	9 00	11 40	1 45	3 30	4 40	4 40	9 30
								▲BOSTON, MA									9 15
...	...	8 15	4 15	4 15	1 00	...	9 15	9 15 Lv Boston, MA (255) Ar	...	...	10 45	6 00	2 15 10 35	...	8 55	5 35 4 45	8 55 1 30	4 45
		12 25	8 20	8 20	4 55		1 55	1 55 Ar Albany, NY	Lv								
12 30	9 15	4 30	3 00	1 00	1 00	9 00	9 00	Lv Boston, MA (109)	Ar	5 40	9 35	12 10	1 55	4 20	5 55	5 55	8 55
↓	10 25	↓	4 10	2 10	2 10	10 10	10 10 Ar	Worcester, MA	Lv					3 10			
2 25	11 45	6 35	5 05	3 30	3 30	11 30	11 30 Ar	Hartford, CT	Lv	3 45	7 30	10 15	11 50	1 50	3 50	3 50	6 50
↓	12 35	7 50	↓	↓	↓	↓	Ar	New Haven, CT	Lv	2 45	6 15	8 50	↑	12 40	↑	↑	↑
5 10	2 50	9 40	7 55	6 20	6 20	2 20	2 20 Ar	NEW YORK, NY (ET)	Lv	12 30	4 15	6 15	9 00	11 00	1 00	1 00	4 00

▲ — Full service agency handling tickets, baggage and express, including C.O.D. express.
🚌 — Package express pickup and delivery service available at this location.
† — Flag stop. Bus will stop on signal to receive and discharge passengers.
≡ — Rest stop.
✕ — Meal or lunch stop.
■ — or HS — Highway stop — does not go into town or agency.
D — or d — Stops only to discharge passengers at agency or in town. Times shown are approximate.

All trips operate daily unless otherwise noted.
Italic type denotes connecting service.
AM — Light face figures.
PM — Bold face figures.

AT — Atlantic Time.
ET — Eastern Time.

FREQUENCY CODES
1 — MONDAY
2 — TUESDAY
3 — WEDNESDAY
4 — THURSDAY
5 — FRIDAY
6 — SATURDAY
7 — SUNDAY
H — HOLIDAY
X — EXCEPT

Shaded trips operate less than daily

EXAMPLE:
X67H EQUALS: Except Saturday, Sunday, Holiday

02002-0730gt

1. What time does bus #635 leave Portland to go south? _____
   What time does it arrive in Boston? _____
   How long does the trip take? _____

2. What other bus leaves Portland to go south the same time as #635?
   _____
   What time does it arrive in Boston? _____ How long does the trip take? _____
   _____ Which bus arrives in Boston earlier? _____ How much earlier?
   _____

3. What is the earliest bus in the day to leave Portsmouth and go north to Portland? _____ What time does it leave Portsmouth? _____
   What time does it arrive in Portland? _____ How long does the trip take? _____

© 1987, 1996 J. Weston Walch, Publisher   *Clock and Calendar Skills*

Board Game: On the Job

Objective

Students record work times and add times to determine 8-hour days and a 40-hour week.

Materials needed

    One copy each of pages 86 and 87 to make board game
    Piece of lightweight cardboard (11" × 16") to back playing board
    Glue
    Buttons (one for each student) to use as playing pieces
    Time card for each player (see page 88)
    Cards (one copy each of pages 89, 90, 91, and 92)
    Scissors
    Die
    Pencil for each student

To prepare the game

1. Cut along the dotted line on page 86. Glue the two gameboard pages to the cardboard.
2. Cut time cards apart. One is needed for each player.
3. Cut apart the four pages of cards. There should be a total of 36 cards.

To play the game

1. Shuffle the cards and place them in "the office."
2. Each player writes his or her name on a time card. Two to four players may play at a time.
3. Each player chooses a button for a playing piece and places it outside the front door.
4. Players roll the die to determine who goes first. Highest roll starts the game.
5. The first player rolls the die and moves the playing piece through the door and the indicated number of squares. Pieces move in clockwise direction around the board.
6. If the player lands on a unit of work ($\frac{1}{2}$ hour, 1 hour, $1\frac{1}{2}$ hours, or 2 hours), he or she writes the time on the time card under Monday.
7. If the player lands on a square that says "Report to the Office," he or she draws a card and does what the card says.
8. When the player is finished, the person to the left takes a turn.
9. Play continues with each person adding time to his or her time card. When at least eight hours have been accumulated under Monday, start writing the work hours under Tuesday.

10. Units of work cannot be split. For example, if a player has 7 hours under Monday and then lands on $1\frac{1}{2}$ hours, the entire $1\frac{1}{2}$ hours must go under Monday. It would make a total of 8 regular hours and $\frac{1}{2}$ hour of overtime. The extra hour cannot be carried over to Tuesday.

11. When one player has worked all 5 days, the game ends.

12. The winner is the person who has worked the highest total number of hours. (This may be someone other than the person who finishes first.)

Variation

To make the game simpler, do not use the cards. Simply write units of work time (1 hour, $1\frac{1}{2}$ hours, etc.) on those spaces on the playing board. Or have player take another turn when they land on one of the "Report to the Office" spaces.

Name _____ Date _____

© 1987, 1996 J. Weston Walch, Publisher   86   Clock and Calendar Skills

Name _____ Date _____

The Office

Board spaces (clockwise from top-left area):

- two hours of work
- Report to the office — THE OFFICE
- one and one-half hours of work
- one hour of work
- one hour of work
- one and one-half hours of work
- two hours of work
- one-half hour of work
- Report to the office — The Boss' Office
- one-half hour of work
- one hour of work
- one-half hour of work

© 1987, 1996 J. Weston Walch, Publisher   87   Clock and Calendar Skills

TIME CARD		Name:		
Monday	Tuesday	Wednesday	Thursday	Friday
Mon Total: ____	Tue Total: ____	Wed Total: ____	Thu Total: ____	Fri Total: ____
Regular: _____ hours	Regular: _____ hours	Regular: _____ hours	Regular: _____ hours	Regular: _____ hours
Overtime: _____ hours	Overtime: _____ hours	Overtime: _____ hours	Overtime: _____ hours	Overtime: _____ hours
Total Hours Worked for the Week: _____				

TIME CARD		Name:		
Monday	Tuesday	Wednesday	Thursday	Friday
Mon Total: ____	Tue Total: ____	Wed Total: ____	Thu Total: ____	Fri Total: ____
Regular: _____ hours	Regular: _____ hours	Regular: _____ hours	Regular: _____ hours	Regular: _____ hours
Overtime: _____ hours	Overtime: _____ hours	Overtime: _____ hours	Overtime: _____ hours	Overtime: _____ hours
Total Hours Worked for the Week: _____				

Car wouldn't start so you are late to work Lose 1 turn	Win good attendance award Add 4 hours to your time card	Work through the coffee break Add ½ hour to your time card
Have to leave work early Lose 1 turn	The suggestion is used that you put in the suggestion box Add 2 hours to your time card	Find easier way to do your job Add 1½ hours to your time card
Stay home with sick sister Lose 2 turns	Volunteer to work Saturday Take 1 extra turn	Customer compliments the boss about your work Add 1 hour to your time card

| Sick with the flu so stay home from work

Lose 2 turns | Thanksgiving bonus

Add 2 hours to your time card | Train a new worker

Add 1½ hours to your time card |
|---|---|---|
| Spill soup on lunchroom floor and stop to clean it up

Lose 1 turn | Boss praises you for good work

Add 1 hour to your time card | Finish your work early

Add ½ hour to your time card |
| Have a headache so go home early

Lose 1 turn | Heat wave so boss lets *everyone* go home early

All players add 1 hour to their time cards | Find lost supplies

Take 1 extra turn |

Accident holds up traffic so you are late to work — Lose 1 turn	Boss in good mood lets you go home early — Add 1 hour to your time card	Flood warnings so boss lets you go home early — Add 2 hours to your time card
Snowstorm so unable to go to work — Lose 2 turns	Equipment breaks down so stay late to make up work — Add 1½ hours to your time card	Work through lunch — Take 2 extra turns
Dentist appointment so you must leave work for an hour — Lose 1 turn	Complete your work early — Add ½ hour to your time card	Arrive at work early — Take 1 extra turn

Careless on the job so cut your hand — Lose 1 turn	Agree to last-minute change in work schedule — Add ½ hour to your time card	Boss comments on your good attitude at work — Add 2 hours to your time card
Electricity goes out during night so alarm doesn't go off and you are late for work — Lose 1 turn	Birthday bonus — Take 1 extra turn	Supervisor gives you rave review on monthly report — Add 1½ hour to your time card
Get in a fight at work so boss sends you home — Lose 2 turns	Holiday tomorrow — Give yourself the total 8 hours on your time card for tomorrow	Find misplaced orders — Add 1 hour to your time card

GAME: THE PRINT SHOP

Objective

Students fill out a time sheet, using tenths of an hour.

Materials needed

Cards—make five copies of page 95 to make a deck of 50 cards for each group of four to five students
Two dice for each group of four to five students
Copy of time sheet (page 96) for each student
Pencil for each student

To prepare the game

1. Prepare a deck of cards for each group of four to five students. Make five copies of page 95. Cut the cards apart to have a deck of 50 cards.
2. Make each deck of cards a different color (for easier sorting).
3. Make one copy of the time sheet (page 96) for each student.

To introduce the game

1. The use of digital clocks makes it very easy to divide the time into tenths of an hour (0.1 hour, 0.2 hour, 0.3 hour, etc.)
2. There are 60 minutes in an hour and so there are 6 minutes in each tenth of an hour. Therefore, 12 (2 × 6) minutes equals 0.2 hour; there are 18 (3 × 6) minutes in 0.3 hour, etc.
3. During your day at the Print Shop, you will make out your time sheet in tenths of an hour. (Pass out time sheets.) If you are doing a collating job that takes you 0.3 of an hour and you start at 7, you work at the job until 7:18. It would be indicated like this on your time sheet:

7:00	
7:06	} collating
7:12	
7:18	
7:24	

Then you would start another job at 7:18.

To play the game

1. Players roll a die to decide who plays first. Highest number starts.
2. The player draws a card which tells what job to do.
3. The player rolls the two dice to see how long the job takes and marks that time on the time sheet.

Roll 2 = 0.2 hour = 12 minutes
Roll 3 = 0.3 hour = 18 minutes
Roll 4 = 0.4 hour = 24 minutes
Roll 5 = 0.5 hour = 30 minutes
Roll 6 = 0.6 hour = 36 minutes
Roll 7 = 0.7 hour = 42 minutes
Roll 8 = 0.8 hour = 48 minutes
Roll 9 = 0.9 hour = 54 minutes
Roll 10 = 1.0 hour = 60 minutes
Roll 11 = 1.1 hour = 66 minutes
Roll 12 = 1.2 hour = 72 minutes

The player discards the card.

4. Play continues with the player to the left.

5. As the game progresses, players draw other cards, roll the dice, and continue listing jobs on the time sheet.

6. The cleanup card is for 0.1 hour. No roll of the dice is required.

7. No roll of the dice is needed for the lunch card either (it is 0.5 hour long), but it can be used only during certain times. The half hour lunch break can be started only between 11:00 and 12:30.

    If it is earlier than 11:00 on the time sheet and the player draws a lunch card, the card must be discarded, and play then continues with next player.

    Players cannot work past 12:30 without starting lunch. If they get a work assignment card that takes them beyond 12:30, they must discard it and hope to draw a lunch card on the next turn.

8. After lunch, job assignments continue until end of the day. The workday ends at 3:30, but players may continue on a job until 4:00. If they draw a card before 3:30 and the roll of the dice shows they would have to work after 4:00, they cannot do the job. They must discard the card and pass the dice to the next player.

9. If there are no more cards to draw and players have not finished the game, the discard pile should be shuffled and reused.

10. If desired, have players total the amount of time they worked on each type of job assignment.

Note

A sample filled-in time sheet is on page 97.

stitching	folding
printing	plastic binding
inserts	hole punching
collating	cutting
lunch (0.5 hour)	cleanup (0.1 hour)

THE PRINT SHOP Time Sheet	Name: Date:	
7:00	10:00	1:00
7:06	10:06	1:06
7:12	10:12	1:12
7:18	10:18	1:18
7:24	10:24	1:24
7:30	10:30	1:30
7:36	10:36	1:36
7:42	10:42	1:42
7:48	10:48	1:48
7:54	10:54	1:54
8:00	11:00	2:00
8:06	11:06	2:06
8:12	11:12	2:12
8:18	11:18	2:18
8:24	11:24	2:24
8:30	11:30	2:30
8:36	11:36	2:36
8:42	11:42	2:42
8:48	11:48	2:48
8:54	11:54	2:54
9:00	12:00	3:00
9:06	12:06	3:06
9:12	12:12	3:12
9:18	12:18	3:18
9:24	12:24	3:24
9:30	12:30	3:30
9:36	12:36	3:36
9:42	12:42	3:42
9:48	12:48	3:48
9:54	12:54	3:54

Total Time		
stitching _____	folding _____	printing _____
plastic binding _____	inserts _____	hole punching _____
collating _____	cutting _____	cleanup _____

THE PRINT SHOP Time Sheet		Name: *George Brown* Date: *5/17*			
7:00 ⌐		10:00		1:00	
7:06	*collating*	10:06		1:06	
7:12 ⌐		10:12	*cutting*	1:12	
7:18 ⌐		10:18		1:18	
7:24		10:24		1:24	
7:30	*hole punching*	10:30 ⌐		1:30	*folding*
7:36		10:36 ⌐		1:36	
7:42		10:42		1:42	
7:48 ⌐		10:48		1:48	
7:54 ⌐		10:54	*hole punching*	1:54	
8:00		11:00		2:00	
8:06		11:06		2:06 ⌐	
8:12	*cutting*	11:12		2:12 —	*cleanup*
8:18		11:18 ⌐		2:18 ⌐	
8:24		11:24 ⌐		2:24	
8:30 ⌐		11:30	*inserts*	2:30	
8:36 —	*cleanup*	11:36		2:36	*inserts*
8:42 ⌐		11:42 ⌐		2:42	
8:48		11:48 ⌐		2:48	
8:54		11:54		2:54 ⌐	
9:00		12:00	*lunch*	3:00 ⌐	
9:06		12:06		3:06	
9:12	*inserts*	12:12 ⌐		3:12	
9:18		12:18 ⌐		3:18	*stitching*
9:24		12:24		3:24	
9:30 ⌐		12:30		3:30	
9:36 ⌐		12:36	*cutting*	3:36 ⌐	
9:42		12:42		3:42	
9:48		12:48		3:48	
9:54		12:54 ⌐		3:54	

Total Time:					
stitching	*0.7 hours*	folding	*1.2 hours*	printing	
plastic binding		inserts	*2.0 hours*	hole punching	*1.4 hours*
collating	*0.3 hours*	cutting	*2.4 hours*	cleanup	*0.2 hours*

© 1987, 1996 J. Weston Walch, Publisher · *Clock and Calendar Skills*

DAYS OF THE WEEK
Directions for Worksheet 28

Objectives

1. Students match days of the week with their abbreviations.
2. Students identify days of the week on a calendar.
3. Students recognize daily relationships such as tomorrow, yesterday, day after tomorrow, day before yesterday, a week from . . .

Materials needed

Copy of Worksheet 28 for each student
Pencil for each student
Calendar for current year
Chalkboard and chalk

Introduce worksheet

1. Before class, make a sketch on the chalkboard of this month's calendar page. Make it large enough for all to see. Include names of days of the week. (Sunday, Monday, etc.)
2. The days on this calendar are arranged in seven columns. What are the days in this first column called? (Sunday) Next column? (Monday) Continue across for Tuesday, etc.
3. If we take one row across—a Sunday, a Monday, etc.—what do we have? (a week)
4. What day of the week is the 3rd of this month? What day of the week will the 7th be? The 12th? The 23rd?
5. How many Fridays will there be this month? How many Sundays? How many Thursdays?
6. What day of the week is the last day of the month?
7. What is the day after Thursday? (Friday) What is the day before Tuesday? (Monday) What is a week from Sunday? (Sunday)
8. Pretend today is Sunday. What day will it be the day after tomorrow? (Tuesday) What day was it the day before yesterday? (Friday)
9. Continue quizzing until students are thoroughly familiar with the days of the week.
10. Students will complete the worksheet by filling in the correct days of the week.

Name _____ Date _____

Reproducible Worksheet 28

Days Word Bank	Abbreviation Word Bank
Friday	Wed
Monday	Mon
Saturday	Tue
Sunday	Fri
Thursday	Sun
Tuesday	Thu
Wednesday	Sat

1. Write the days of the week in order, beginning with Sunday. Beside each day, write its abbreviation.

Day of the Week	Abbreviation

2. Look at a calendar showing this month. What day of the week is the

   1st? _____ 13th? _____ 10th? _____
   18th? _____ 4th? _____ 26th? _____
   23rd? _____ 14th? _____ 7th? _____

3. What day of the week is Christmas this year? _____
4. What day of the week is your birthday this year? _____
5. What day of the week is today? _____
6. What day of the week was the day before yesterday? _____
7. What day of the week will it be tomorrow? _____
8. What day will it be a week from day after tomorrow? _____

WEEKLY SCHEDULE
Directions for Worksheet 29

Objective

    Students identify activities unique to each day of the week.

Materials needed

    Copy of Worksheet 29 for each student
    Pencil for each student
    Chalkboard and chalk

Introduce worksheet

1. Before class, make five columns and head them as shown below:

Monday	Tuesday	Wednesday	Thursday	Friday

2. These five columns show the five days we are together in school. What things do we do every day at school? (eat lunch, have classes, etc.)

3. Now, instead of the things we do every day, we want to talk about the things that make each day special. What do we do on Monday that we don't do on Tuesday? or what makes Tuesday special? Do you go to the library one day? or have art class on one day? Write the special things about each day in the correct column.

4. This listing has five columns because there are five days of school each week. If you made a list of special things about each day of the week, how many columns would you need? (seven)

5. Students will complete the worksheet by writing things that make each day special in the appropriate columns.

Name _____   Date _____

Reproducible Worksheet 29

Some people do certain things on certain days of the week. Look at the following list:

Laundry	Go to church	Sleep late	Grocery shopping
Prepare dinner	Music class	Clean room	Dance lesson
Take out trash	Exercise	Baby-sit	Watch favorite TV program
Go to school	Do dishes	Mow lawn	Get allowance/get paid

Which things on the list do you usually do on certain days? Write them in the columns below.

Sunday	Monday	Tuesday	Wednes-day	Thursday	Friday	Saturday

What other activities can you add to any of the days? Do you usually visit a relative on one day of the week? Does someone regularly come to your home for Sunday dinner?

Game: A Busy Week

Objective

    Students schedule a variety of activities to complete a weekly calendar.

Materials needed

    One copy of the weekly calendar (page 103) for each student
    Pencil for each student
    Cards (one copy each of pages 104 through 113)
      Note: The last four cards on page 112 and all cards on page 113 are used only in the more challenging version of game.
    Scissors

To prepare the game

1. Make a copy of each page of cards.
2. Cut the cards apart. There will be 84 cards for the easier game—98 cards in all, including those used in the more challenging version of the game.

To play the game

1. Two to four players may play the game at a time.
2. Shuffle the cards and place in the center of the table.
3. Each player needs a copy of the weekly calendar and a pencil.
4. The first player draws a card. The player enters the activity written on the card in the appropriate time period on his or her weekly calendar.
5. The card is then put to the side and removed from play.
6. Play continues with the player to the left.
7. After several rounds of play, a player may draw a card that has an activity for a time period that player has already filled.
8. If the player is unable to use a card, the player puts the card face up beside the deck of cards.
9. The next player can either draw a card from the deck or use the top card in the discard pile.
10. If the cards in the deck run out before the game is over, shuffle the cards in the discard pile and replace the deck.
11. The first player to fill all 21 spaces in the weekly calendar with an activity is the winner.

Variation

    Add the last 14 cards (on pages 112 and 113) to make the game more challenging. These cards include wild cards which players can use wherever they have spaces. There are also cards that will require players to erase activities that had been scheduled. Some of the cards have activities that will fill two or more spaces. If any of the time periods used in one of these cards has already been filled by a previously selected activity, the new activity cannot be used.

| Name _____ | | Date _____ |

Sunday	morning	
	afternoon	
	evening	
Monday	morning	
	afternoon	
	evening	
Tuesday	morning	
	afternoon	
	evening	
Wednesday	morning	
	afternoon	
	evening	
Thursday	morning	
	afternoon	
	evening	
Friday	morning	
	afternoon	
	evening	
Saturday	morning	
	afternoon	
	evening	

© 1987, 1996 J. Weston Walch, Publisher

Clock and Calendar Skills

Jog 5 miles Sunday morning	Go to church Sunday, 10 A.M.
Read in bed until noon Sunday morning	Plan to sleep until noon Sunday morning
Go bicycling with a friend Sunday afternoon	Deliver flowers to shut-ins Sunday afternoon
Wash windows Sunday afternoon	Picnic in park with a friend Sunday, 1 P.M.
Swim at the pool Sunday, 7:30 P.M.	Do homework Sunday evening

Go to grandparents' for dinner Sunday, 5:30 P.M.	Take little brother to movie Sunday, 6 P.M.
Dentist appointment Monday, 11 A.M.	Clean closet Monday morning
Take dog to vet Monday, 9:30 A.M.	Appointment to get haircut Monday, 11:30 A.M.
Balance checkbook Monday afternoon	Meet friend for lunch Monday, 12:30 P.M.
Mail package at post office Monday afternoon	Shop for sister's birthday gift Monday afternoon

Dinner out with a friend Monday, 6 P.M.	Family conference Monday, 8 P.M.
Friends come to play cards Monday, 7:30 P.M.	Work on hobby Monday evening
Visit friend in hospital Tuesday morning	Get haircut Tuesday, 10 A.M.
Meet friend on coffee break Tuesday, 10:30 P.M.	Write letters Tuesday morning
Tutoring Tuesday, 4 P.M.	Music lesson Tuesday, 3:30 P.M.

| Volunteer at library | Visit grandmother in nursing home |
| Tuesday 1–4 | Tuesday afternoon |

| Schedule evening home alone | Chorus practice |
| Tuesday evening | Tuesday, 7 P.M. |

| Pottery Class | Go to a movie |
| Tuesday, 7 P.M. | Tuesday, 7:30 P.M. |

| Bake cake for birthday party | Job interview |
| Wednesday morning | Wednesday, 10 A.M. |

| Mow lawn | Return books to library |
| Wednesday morning | Wednesday morning |

Take bicycle to be repaired Wednesday afternoon	Counseling appointment Wednesday, 2:30 P.M.
Buy new shoes Wednesday afternoon	Pick up order at co-op Wednesday afternoon
Watch favorite TV program Wednesday, 8 P.M.	Go roller-skating Wednesday, 7:30 P.M.
Scout meeting Wednesday, 7 P.M.	Church service Wednesday, 7:30 P.M.
Organize cellar Thursday morning	Take shoes to cobbler for repair Thursday morning

Grocery shopping Thursday morning	Pick up bread at the bakery Thursday morning
Doctor's appointment Thursday, 2 P.M.	Do laundry Thursday afternoon
Art class Thursday, 4 P.M.	Family photograph taken Thursday, 3 P.M.
Go bowling with a friend Thursday, 7 P.M.	Dinner at favorite restaurant Thursday, 7:30 P.M.
Shop for a new living room chair Thursday evening	Housewarming for a friend Thursday, 7 P.M.

Clothes shopping with a friend Friday morning	Rake the yard Friday morning
Go to pick up paycheck Friday morning	Go to friend's house for coffee Friday, 10:30 A.M.
Clean attic Friday afternoon	Basketball practice Friday, 3 P.M.
Volunteer at hospital Friday afternoon	Job interview Friday, 2 P.M.
Baby-sit Friday, 6:30 P.M.	Sleep over at a friend's house Friday evening

Go to a party Friday, 8 P.M.	Go to stock car races Friday, 8 P.M.
Wash kitchen floor Saturday morning	Stack firewood Saturday morning
Take TV for repairs Saturday morning	Take rubbish to the dump Saturday morning
Wedding Saturday, 1 P.M.	Help neighbor paint house Saturday afternoon
Weed the garden Saturday afternoon	Go to baseball game Saturday, 1:30 P.M.

Wait tables at public supper Saturday, 5 P.M.	Go dancing Saturday, 8 P.M.
Go to bed early (by 6 P.M.) Saturday evening	Potluck supper at a friend's house Saturday, 6 P.M.

The following cards on this and the next page
may be added to the game for additional challenge.

Emergency in family Cancel plans for Monday afternoon and evening	Plans for Thursday evening canceled
Monday morning plans canceled	Invite friend for dinner Any free night

Personal day Any free day morning, afternoon, and evening — **WILD CARD**	Shopping spree Any free afternoon — **WILD CARD**
Trip to state park with family Leave: Sunday, 8 A.M. Return: Sunday near midnight	Practice piano Monday, Wednesday, and Friday afternoons
Aunt and Uncle come for dinner Prepare dinner: Monday afternoon Entertain: Monday evening	Deliver newspapers for a friend Monday, Tuesday, Wednesday mornings
Play rehearsal Tuesday, Wednesday, Thursday afternoons	Studying hard for exam Stay home Tuesday, Wednesday, Thursday evenings
Go on an overnight with Scouts Leave: Friday, 5 P.M. Return: Saturday, 9 P.M.	Clean out garage Friday morning and afternoon — **WILD CARD**

MONTHS OF THE YEAR
Directions for Worksheet 30

Objectives
1. Students match months of the year with their abbreviations.
2. Students write dates using full names of the months, using abbreviations for the months, and using numerical representations for the months.

Materials needed
Chalkboard and chalk
Copy of Worksheet 30 for each student
Pencil for each student

Introduce worksheet
1. Ask one student in which month his or her birthday is. Write the full name of the month on the board. Ask for another month when someone has a birthday. Continue until you have all the months with birthdays. Then write the months that are missing until all 12 months are on the board.
2. Let's come up with a shorter way to write the names of the months. How about if we just use the first letter for each month. Go down through all 12 months and write the first letter of each month beside it.
3. Will this work? (no) Why not? (Some of the months have the same letters—you can't tell April from August, January from July or June, March from May.)
4. Let's add the second letter for each month.
5. Will this work? (no) Why not? (You still can't tell March from May.)
6. So let's add a third letter—it's still a lot shorter to write Feb than to write February. It's easier to spell too!
7. Will this work? (yes)
8. What month will be the same abbreviated as it is spelled out? (May)
9. Now let's put these 12 months in order. What's the first month of the year? (January) Write a "1" beside January. Continue with "2" for February, "3" for March, etc.
10. Let's look at today's date. Write the date on the board using the full name of the month.
11. How could we write this same date more easily? (Use abbreviation for month.) Write date again using abbreviation.
12. Now let's write this date using just numbers. What number is the month? Write date on board in numbers, using dashes to separate month, date, and year.
13. Write another date on the board, using the full name of the month. Ask one student to write the date with the abbreviation of the month. Have another write the date using using only numbers.
14. Have a student write another date on the board and have other students write the same date in other ways. Continue until students can easily write and recognize dates written in all three ways.

Name _____   Date _____

Reproducible Worksheet 30

Month Word Bank		Abbreviation Word Bank	
April	June	Oct	Feb
August	March	Mar	Apr
December	May	Jan	Sept
February	November	Aug	May
January	October	Nov	Jul
July	September	Jun	Dec

Write the names of the months in order. Beside each month, write its abbreviation.

Name of Month   **Abbreviation**

1. _____ _____
2. _____ _____
3. _____ _____
4. _____ _____
5. _____ _____
6. _____ _____
7. _____ _____
8. _____ _____
9. _____ _____
10. _____ _____
11. _____ _____
12. _____ _____

Write the following dates in three ways. Use the full name of the month. Use the abbreviation for the month. Use the number for the month. The first is done for you.

January 17, 1921	Jan 17, 1921	1-17-21
	Apr 23, 1956	
December 9, 1983		
		7-6-38
	Mar 30, 1993	
		10-31-75

© 1987, 1996 J. Weston Walch, Publisher   *Clock and Calendar Skills*

WRITING DATES WITH NUMBERS
Directions for Worksheet 31

Objectives
1. Students use numbers to write dates.
2. Students practice filling in forms.

Materials needed
Copies of Worksheet 31 for each student
Pencil for each student
Chalkboard and chalk

Introduce worksheet
1. Before class, make a few of the following boxes on the chalkboard. Make several of each type.

2. Write the date "October 22, 1948" on the chalkboard beside one of the longer blocks. How can this long date be written in these six squares? (Write very small.)
3. The number in the middle is easy. There are two digits, so let's just write them in the middle.
4. How about the year? Can we use just part of it? If we use just the last two digits, it will fit. (Write "48" in last two boxes.)
5. How can we change October to a number? If January is 1, February is 2, etc., what number will October be? (10) So let's write a 10 in the first squares to represent the 10th month—October.
6. Let's try another date. December 25, 1993. How would that be written? (12-25-93)
7. What about this date—March 30, 1950. The "30" is easy. So is the year—"50." What number represents March? (3) Leave the first block empty and put the "3" in the second block.
8. How about January 3, 1932? (Leave the first square blank, then a "1" for January. Leave the third square blank, then a "3." Write "32" for the year.)
9. Sometimes we don't need to include the specific date. Just the month and year is all that is wanted. What month did you start school? Include the month and the year and write the date in one of these blocks with four squares.
10. Continue until students can easily rewrite dates using numbers.
11. Students will complete the worksheet by filling out the forms.

Name _____ Date ☐☐–☐☐–☐☐

Reproducible Worksheet 31

1. List the members of your family, oldest to youngest.

   Last Name   First Name   MI   Birthday Month–Day–Year

2. Where have you lived? List present address first.

   Street   Town or City   State   Month–Year to Month–Year

3. Schools attended. List present school first.

   Name of School   Town or City   Month–Year to Month–Year

© 1987, 1996 J. Weston Walch, Publisher

Clock and Calendar Skills

CALENDAR

Objective
Make individual calendars for students in the class.

Materials needed
A current calendar
Scissors
Transparent tape
Scraps of white paper
Words and numbers on the next page
Calendar outlines on the following pages

Introduce worksheet
1. Make a copy of the words and numbers on the following page.
2. Consult a current calendar to determine which day of the week is January 1.
3. Make a copy of the calendar outline that starts on that day of the week.
4. Cut out the word "January" and the appropriate year. Tape these in the places indicated on the calendar outline.
5. Make the necessary number of copies of January.
6. Consult a current calendar to determine which day of the week is February 1.
7. Make a copy of the calendar outline that starts on that day of the week.
8. Since there are not 31 days in February, tape small scraps of paper to cover the unnecessary numbers and dividing lines.
9. Make the necessary number of copies of February.
10. Continue in this manner for each of the 12 months.
11. Worksheets 32–35 provide activities to complete with the calendars.

Variations
1. If it is near the end of the year, it might be better to make the school year calendar or a calendar for the next year.
2. Students can attach the calendar pages to larger pieces of paper and draw illustrations to accompany each month.

JANUARY february

MARCH April

MAY JUNE JULY

AUGUST SEPTEMBER

October NOVEMBER

december

1996 1997 1998 1999

2000 2001 2002 2003

2004 2005 2006 2007

January	February	March	April

May	**June**	**July**	**August**

September	**October**	**November**	**December**

MONTH _____ YEAR _____

SUNDAY	MONDAY	TUESDAY	WEDNESDAY	THURSDAY	FRIDAY	SATURDAY
1	2	3	4	5	6	7
8	9	10	11	12	13	14
15	16	17	18	19	20	21
22	23	24	25	26	27	28
29	30	31				

MONTH _____ YEAR _____

SUNDAY	MONDAY	TUESDAY	WEDNESDAY	THURSDAY	FRIDAY	SATURDAY
	1	2	3	4	5	6
7	8	9	10	11	12	13
14	15	16	17	18	19	20
21	22	23	24	25	26	27
28	29	30	31			

MONTH _____   YEAR _____

SUNDAY	MONDAY	TUESDAY	WEDNESDAY	THURSDAY	FRIDAY	SATURDAY
		1	2	3	4	5
6	7	8	9	10	11	12
13	14	15	16	17	18	19
20	21	22	23	24	25	26
27	28	29	30	31		

Clock and Calendar Skills

MONTH _____   YEAR _____

SUNDAY	MONDAY	TUESDAY	WEDNESDAY	THURSDAY	FRIDAY	SATURDAY
			1	2	3	4
5	6	7	8	9	10	11
12	13	14	15	16	17	18
19	20	21	22	23	24	25
26	27	28	29	30	31	

MONTH _____ YEAR _____

SUNDAY	MONDAY	TUESDAY	WEDNESDAY	THURSDAY	FRIDAY	SATURDAY
				1	2	3
4	5	6	7	8	9	10
11	12	13	14	15	16	17
18	19	20	21	22	23	24
25	26	27	28	29	30	31

MONTH   YEAR

SUNDAY	MONDAY	TUESDAY	WEDNESDAY	THURSDAY	FRIDAY	SATURDAY
					1	2
3	4	5	6	7	8	9
10	11	12	13	14	15	16
17	18	19	20	21	22	23
24	25	26	27	28	29	30
31						

MONTH _____ YEAR _____

SUNDAY	MONDAY	TUESDAY	WEDNESDAY	THURSDAY	FRIDAY	SATURDAY
						1
2	3	4	5	6	7	8
9	10	11	12	13	14	15
16	17	18	19	20	21	22
23	24	25	26	27	28	29
30	31					

HOLIDAYS ON SAME DATE
Directions for Worksheet 32

Objectives

1. Students list holidays that occur on the same date each year.

2. Students use a calendar to locate when we observe specific holidays that occur on the same date each year.

Materials needed

Individual student calendars (see page 118)
Copy of Worksheet 32 for each student
Pencil for each student
Chalkboard and chalk
Scissors (optional)
Glue (optional)

Introduce worksheet

1. Ask students to name holidays they celebrate.

2. Write the names of the holidays on the chalkboard as they name them.

3. Circle the holidays that occur on the same date each year (New Year's Day, Valentine's Day, Halloween, etc.) These are the holidays we are going to talk about today. They happen on the same date each year.

4. Students will complete the worksheet by locating the holidays on their calendars.

Variation

If students have trouble writing, they can cut out the pictures that illustrate each holiday and glue them on the appropriate dates.

Name _____   Date _____

Reproducible Worksheet 32

Some holidays are on the same date each year. Find these holidays on your calendar.

New Year's Day

January 1

Groundhog Day

February 2

Valentine's Day

February 14

St. Patrick's Day

March 17

All Fools' Day

April 1

Flag Day

June 14

Independence Day

July 4

Halloween

October 31

Veterans Day

November 11

Christmas

December 25

© 1987, 1996 J. Weston Walch, Publisher   129   Clock and Calendar Skills

HOLIDAYS ON SAME DAY
Directions for Worksheet 33

Objectives

1. Students list holidays that occur on the same day of the week each year.

2. Students use a calendar to locate specific holidays that occur on the same day of the week each year.

Materials needed

Individual student calendars (see page 118)
Copy of Worksheet 33 for each student
Pencil for each student
Scissors (optional)
Glue (optional)

Introduce Worksheet

1. Ask students what they know about Thanksgiving. (eat turkey; Pilgrims, etc.) How about when we celebrate it? (November—on a Thursday) It doesn't have a specific date the way Christmas does. Instead, Thanksgiving is always on the fourth Thursday of November.

2. There are other holidays that are celebrated on the same day of the week each year.

3. Students will complete the worksheet by locating the holidays on their calendars.

Variation

If students have trouble writing, they can cut out the pictures that illustrate each holiday and glue them on the appropriate dates.

Reproducible Worksheet 33

Some holidays are on the same day of the week each year. Find these holidays on your calendar.

Martin Luther King's Birthday

3rd Monday in January

Presidents' Day

3rd Monday in February

Mother's Day

2nd Sunday in May

Memorial Day

4th Monday in May

Father's Day

3rd Sunday in June

Labor Day

1st Monday in September

Grandparents' Day

1st Sunday in September

Columbus Day

2nd Monday in October

Election Day

1st Tuesday in November

Thanksgiving

4th Thursday in November

PERSONAL HOLIDAYS
Directions for Worksheet 34

Objectives

1. Students list their own personal days of celebration.
2. Students use a calendar to locate their own personal days of celebration.

Materials needed

Individual student calendars (see page 118)
Copy of Worksheet 34 for each student
Pencil for each student
Chalkboard and chalk
List of birthdays of members of the class

Introduce worksheet

1. Write the date of one student's birthday on the chalkboard. Ask the class if this is an important date to anyone. When someone acknowledges the date, ask why it is a special day.
2. Write another birthday on the board for a student to identify.
3. Each of us has important dates that we celebrate every year. These dates are not always important to other people.
4. Students will complete the worksheet by listing days that are important to them and then locating those dates on a calendar.

Name _____   Date _____

Reproducible Worksheet 34

Each person has special days and events to remember and celebrate. In the space below, write the dates of your own celebrations. Then find these dates on our calendar.

EVENT   DATE

Your Birthday: _____

Birthdays of Other Family Members:

_____

_____

_____

Other Birthdays to Remember:

_____

_____

_____

Anniversaries:

_____

_____

_____

Your Vacation: _____

Religious Holidays (Easter, Hanukkah, etc.):

_____

_____

Other: _____

THE SEASONS
Directions for Worksheet 35

Objectives

1. Students identify the four seasons of the year and list characteristic weather for each season.

2. Students use a calendar to locate the first day of each season and identify which months are included in each season.

Materials needed

Individual student calendars (see page 118)
Copy of Worksheet 35 for each student
Pencil for each student
Chalkboard and chalk

Introduce worksheet

1. How many different kinds of weather conditions can we list on the chalkboard? (Write on board as students suggest, rain, sunny, snow, cold, hot, windy, freezing, tornado, etc.)

2. What kind of weather might it be likely for us to have here this week? (Circle the answers that the students give.)

3. Talk about the other kinds of weather that are not likely this week. When might those weather conditions exist?

4. Different seasons of the year have different weather conditions.

5. Students will complete the worksheet by locating the different seasons on their calendars and identifying weather conditions during each season.

Name _____ Date _____

Reproducible Worksheet 35

The first day of spring is March 20.
Find this date on your calendar.

The first day of autumn is September 22.
Find this date on your calendar.

The first day of summer is June 21.

Find this date on your calendar.

The first day of winter is December 21.

Find this date on your calendar.

What months are part of spring? _____
_____

What is the weather like in spring? _____
_____

What months are part of summer? _____
_____

What is the weather like in summer? _____
_____

What months are part of autumn? _____
_____

What is the weather like in autumn? _____
_____

What months are part of winter? _____
_____

What is the weather like in winter? _____
_____

SEASONAL ACTIVITIES
Directions for Worksheet 36

Objective

    Students identify seasons when different activities take place.

Materials needed

    Copy of Worksheet 36 for each student
    Blue, green, red, and black markers for each student

Introduce worksheet

1. Tell students that you have an idea for what to do on Saturday—plant the flower garden in the morning and then go skating on the lake.

2. When some respond to the unlikelihood of doing those two activities on the same day, suggest instead that you go cross-country skiing and then relax by going sailing.

3. Some activities are seasonal. In some areas of the world even more than others, the different seasons bring different weather conditions and different activities.

4. Students will complete the worksheet by making a blue check next to spring activities, a green check next to summer activities, a red check next to fall activities, and a black check next to winter activities.

Name _____ Date _____

Worksheet 36

What activities take place in the different seasons?

    Put a blue ✔ next to spring activities.
    Put a green ✔ next to things that are done in the summer.
    Put a red ✔ next to activities done in the fall.
    Put a black ✔ beside winter activities

Some activities may take place in several seasons. They will receive several checks.

☐ mow the lawn   ☐ ride a bicycle   ☐ celebrate Halloween   ☐ swim in the ocean

☐ plant the garden   ☐ cross-country ski   ☐ put away winter clothes   ☐ celebrate Thanksgiving

☐ ice skate on the lake   ☐ rake the leaves   ☐ make a snowman   ☐ picnic in the park

☐ play baseball   ☐ celebrate Christmas   ☐ start school year   ☐ celebrate Easter

© 1987, 1996 J. Weston Walch, Publisher   *Clock and Calendar Skills*